Fundamentals of Marine Acoustics

FURTHER TITLES IN THIS SERIES

1 J.L. MERO
THE MINERAL RESOURCES OF THE SEA
2 L.M. FOMIN
THE DYNAMIC METHOD IN OCEANOGRAPHY
3 E.J.F. WOOD
MICROBIOLOGY OF OCEANS AND ESTUARIES
4 G. NEUMANN
OCEAN CURRENTS
5 N.G. JERLOV
OPTICAL OCEANOGRAPHY
6 V. VACQUIER
GEOMAGNETISM IN MARINE GEOLOGY
7 W.J. WALLACE
THE DEVELOPMENT OF THE CHLORINITY/SALINITY CONCEPT IN OCEANOGRAPHY
8 E. LISITZIN
SEA-LEVEL CHANGES
9 R.H. PARKER
THE STUDY OF BENTHIC COMMUNITIES
10 J.C.J. NIHOUL (Editor)
MODELLING OF MARINE SYSTEMS
11 O.I. MAMAYEV
TEMPERATURE—SALINITY ANALYSIS OF WORLD OCEAN WATERS
12 E.J. FERGUSON WOOD and R.E. JOHANNES
TROPICAL MARINE POLLUTION
13 E. STEEMANN NIELSEN
MARINE PHOTOSYNTHESIS
14 N.G. JERLOV
MARINE OPTICS
15 G.P. GLASBY
MARINE MANGANESE DEPOSITS
16 V.M. KAMENKOVICH
FUNDAMENTALS OF OCEAN DYNAMICS
17 R.A. GEYER
SUBMERSIBLES AND THEIR USE IN OCEANOGRAPHY AND OCEAN ENGINEERING
18 J.W. CARUTHERS
FUNDAMENTALS OF MARINE ACOUSTICS
19 J.C.J. NIHOUL (Editor)
BOTTOM TURBULENCE
20 P.H. LEBLOND and L.A. MYSAK
WAVES IN THE OCEAN
21 C.C. VON DER BORCH (Editor)
SYNTHESIS OF DEEP-SEA DRILLING RESULTS IN THE INDIAN OCEAN
22 P. DEHLINGER
MARINE GRAVITY
23 J.C.J. NIHOUL
HYDRODYNAMICS OF ESTUARIES AND FJORDS
24 F.T. BANNER, M.B. COLLINS and K.S. MASSIE (Editors)
THE NORTH-WEST EUROPEAN SHELF SEAS: THE SEA BED AND THE SEA IN MOTION

Elsevier Oceanography Series, 18

Fundamentals of Marine Acoustics

by

JERALD W. CARUTHERS

ELSEVIER SCIENTIFIC PUBLISHING COMPANY
Amsterdam — Oxford — New York 1977

ELSEVIER SCIENTIFIC PUBLISHING COMPANY
335 Jan van Galenstraat
P.O. Box 211, 1000 AE Amsterdam, The Netherlands

Distributors for the United States and Canada:

ELSEVIER/NORTH-HOLLAND INC.
52, Vanderbilt Avenue
New York, N.Y. 10017

First edition 1977
Second impression 1979

Library of Congress Cataloging in Publication Data

Caruthers, Jerald W
Fundamentals of marine acoustics.

(Elsevier oceanography series ; 18)
Bibliography: p.
Includes index.
1. Underwater acoustics. I. Title.
QC242.2.C37 551.4'601 77-4044
ISBN 0-444-41552-1

ISBN 0-444-41552-1 (Vol. 18)
ISBN 0-444-41623-4 (Series)

Printed in The Netherlands

PREFACE

This book is intended for those graduate and upper-level undergraduate students in technical fields who wish to know something about how sound propagates in the ocean. It is also useful for the practicing engineer and scientist. It is based on the contents of a graduate course taught several years in the Department of Oceanography of Texas A & M University and a short course held there two summers. The courses were developed under the auspices of the Sea Grant Program of the National Oceanic and Atmospheric Administration, United States Department of Commerce. Consistent with the philosophy of the Sea Grant Program, the courses emphasized civilian applications and environmental factors. The book was originally published as a Texas A & M University Sea Grant Program report which has been used in several university courses in both the United States and abroad.

I greatly appreciate the support I received from Texas A & M University and the U.S. Sea Grant Program in developing the courses. I am also grateful to the students who helped me compile some of the material and to Mrs. Barbara Webb and Mr. Domenik Summa for drawing the figures.

August 1976 JERALD W. CARUTHERS

CONTENTS

CHAPTER 1

INTRODUCTION

Acoustics is the engineering science that deals with the generation, propagation, and reception of energy in the form of vibrational waves in a material medium. For this energy to be of any value it must contain some form of information; so, generally, understanding the intelligence contained in the transmitted energy is included as a branch of acoustics.

The recorded history of sound in the sea dates to 1490 when Leonardo da Vinci wrote: "If you cause your ship to stop, and place the head of a long tube in the water and place the outer extremity to your ear, you will hear ships at a great distance from you." This is the first example of the passive underwater listening device. Since that day, the engineering science of sound in the sea has come a long way.

Practical uses of underwater sound began just after the turn of the century: The underwater bell was used for a beacon in dangerous waters and for a navigation aid in shallow waters; echo-sounders were used for depth determination; and echo-ranging sonars (then called "asdic") were used for submarine and iceberg detection and location.

Today, bathymetry is determined solely with echo-sounders and the detection of submarines is accomplished primarily with sonars. Echo-ranging systems are also used for the detection, location, and, to some extent, classification of fish shoals. Offshore exploration for petroleum is quite dependent upon the use of subbottom echo-sounding devices. Passive listening and communication devices, as well as acoustic markers and positioning systems, have many applications. More exotic devices such as those for acoustical imaging and holography, as well as for acoustical underwater television telemetry, are in the offing, if not already upon us.

In fact, the use of underwater sound has proliferated to the point that future use may require regulation.*

1.1. SOUND IN THE SEA

The basic concept of sound transmission is fairly simple: Sound is a propagated mechanical disturbance in a medium. Associated with this

* *Present and Future Civil Uses of Underwater Sound*, National Academy of Sciences (Washington, D.C., 1970).

disturbance are pressure and density fluctuations brought about by particle motions. There are basic relations between forces distorting the medium and these particle motions. According to the simplest model, good for small-amplitude waves, the equations of motion, continuity, and state may be combined into a partial differential equation known as a wave equation, e.g.:

$$\nabla^2 p - \frac{1}{c^2}\frac{\partial^2 p}{\partial t^2} = 0 \tag{1.1}$$

where $p = p(x_1, x_2, x_3, t)$ is the excess pressure, $c = c(x_1, x_2, x_3)$ is the speed of sound, and ∇^2 is the Laplacian operator.* Naturally, a specific problem would also include boundary conditions.**

For even the most elementary understanding of sound in the sea, it is necessary to have at least a basic understanding of the nature of sound field. The simplest description of this sound field is the solution of the wave equation in one (spatial) dimension in terms of a sinusoidal plane wave represented by:

$$p(x, t) = p_0 \sin\left(\frac{2\pi x}{\lambda} - 2\pi f t + \phi\right) \tag{1.2}$$

where p_0 is the amplitude of the pressure disturbance, λ the wavelength, f the frequency, and ϕ the phase shift. T, the period, is equal to $1/f$. The argument of the sine function is called the phase. The quantity $f\lambda$ can be shown to be the phase speed c, i.e., $c = f\lambda$.

Keeping in mind that the primary function of acoustic transmissions in the sea is to transmit some form of information, a function to which no other physical phenomenon has been suitably adapted, it is worthwhile to consider some of the fundamental points which control and limit its transmission as compared with various electromagnetic telemetry systems used in air (radio, radar, eye, etc.).

— The velocity of sound in water is small compared to the velocity of electromagnetic waves in air (by a factor of $2 \cdot 10^5$), so the scan rate is greatly reduced.

— High frequencies are strongly attenuated (useful frequencies are generally below 15 kHz), so the data rate is much lower.

— Wavelengths are longer (greater than 0.1 m because $\lambda = c/f$), so the spatial resolution is much lower.

— The gradients in the speed of sound in water are much stronger than gradients of the speed of light in air, thus giving stronger refraction of sound (radii of curvature for rays in the order of tens of miles), so that bearing information is less reliable.

* (x_1, x_2, x_3) are spatial coordinates, e.g., the rectangular coordinates (x, y, z).

** A more complete treatment of the theory of sound propagation is given in Chapter 6. All of the formulas given here are developed in that chapter.

— Background noise and reverberation in the sea are greater, stronger frequency spreading occurs, and the Q (see next chapter) is lower, so detection is more severely limited.

— Water cannot support pressures below a certain minimum, so the amount of energy that may be put into the water is limited by the size, and depends upon the depth of the projector.

Nearly all research and development in underwater acoustics can be traced to one of the above problems. Throughout this book the simple fact that the primary function of underwater acoustics is to transmit information will be stressed and the above problems will constantly appear.

1.2. THE NATURE OF THE ACOUSTIC FIELD

We have already pointed out that the acoustic field may be described by a pressure variation $p(x, y, z, t)$ and that, in certain simple cases, this pressure variation may be represented by a sinusoidal plane wave:

$$p(x, t) = p_0 \sin (kx - \omega t + \phi)$$

where the angular wave number (k) and the angular frequency (ω) have been defined by $k = 2\pi/\lambda$ and $\omega = 2\pi f$.* It is easily shown that the velocity ($c = f\lambda$) is given by $c = \omega/k$.

The general plane wave in one dimension is described by any function of the form:

$$p(x, t) = f(kx - \omega t) \tag{1.3}$$

Any such plane wave may be formed from a sum (or an integral) of appropriate sinusoidal plane waves. For example, a plane wave may be:

$$p(x, t) = \sum_{i=1}^{n} p_i \sin (k_i x - \omega_i t + \phi_i) \tag{1.4}$$

where n may be any number and p_i is the amplitude of the component wave having angular frequency ω_i, wave number k_i, and phase shift ϕ_i. The velocity of each component wave is given by $c_i = \omega_i/k_i$. If the velocities of all the components are the same, the medium is said to be non-dispersive, otherwise it is dispersive. For all practical purposes water is non-dispersive to acoustic waves.

For many applications, the average intensity of the acoustic field is more

---

* The three-dimensional sinusoidal plane wave may be obtained by replacing kx by $\boldsymbol{k} \cdot \boldsymbol{r}$ where $\boldsymbol{k}$ is the wave or propagation vector with magnitude $2\pi/\lambda$ and is in the direction of wave propagation, and $\boldsymbol{r}$ is the position vector of the point in question from an arbitrary origin. But, a coordinate system may always be found in which any plane wave will appear in a one-dimensional form.

useful and more readily measured than the pressure. Intensity is a measure of the rate of energy flow (power) through a unit area perpendicular to the direction of wave propagation:

Intensity = Power/Area.

If the wave is not a plane wave or if there are multiple beams, the situation is more complex and will not be discussed here.

The symbol I and the word "intensity" will generally mean "average intensity" and may be calculated from:

$$I = K\overline{pu} \tag{1.5}$$

where K is a constant and depends upon the choice of units, u is the particle speed, and the bar means time average over an interval τ.

The time average $\bar{X}(t, \tau)$ of any quantity $X(t)$ may be obtained from:

$$\bar{X}(t, \tau) = \frac{1}{\tau}\int_t^{t+\tau} X(t')\mathrm{d}t' \; * \tag{1.6}$$

Note that $\overline{pu} \neq \bar{p}\,\bar{u}$ because p and u are not independent, as will be seen next.

The equations of motion used in developing the wave equation are also useful as they are; for example, Newton's Law states:

$$\mathrm{grad}\, p = -\rho\,\frac{\partial u}{\partial t} \tag{1.7}$$

where u is the particle velocity and ρ is the equilibrium density. For plane waves, the gradient may be taken and the integration performed to get:

$$p = \rho c u \tag{1.8}$$

If this is substituted into equation 1.5, we find:

$$I(t, \tau) = K\frac{\overline{p}^2(t, \tau)}{\rho c}$$

for any acoustic plane wave. As is common for underwater acoustics, pressure is in dyn/cm^2 (microbar, μbar)**, density is in g/cm^3, the speed of

---

* If the signal $X(t)$ is stationary and if τ is large enough, then:

$$\bar{X}(\tau) = \frac{1}{\tau}\int_0^{\tau} X(t')\mathrm{d}t' \tag{1.6a}$$

that is, the average is independent of when it is taken.

** In 1968 the micropascal (10^{-6} newton/meter^2) became the American standard pressure. However, the microbar is still common and will be used in this introduction. 10^{-5} microbar = 1 micropascal.

sound (as used in this formula) is in cm/sec, and intensity is to be watt/cm^2. Thus, K is 10^{-7} (watt/cm^2)/(erg/sec. cm^2) and intensity is given by the practical formula:

$$I = \frac{\overline{p}^2}{\rho c} \cdot 10^{-7} \qquad (1.9)$$

The product of ρc is about 154,000 g/sec. cm^2 for water, so that

$$I = \frac{\overline{p}^2}{154{,}000} \cdot 10^{-7} = 0.65 \cdot 10^{-12} \overline{p}^2 = 0.65 \cdot 10^{-12} p_{rms}^2{}^{*} \qquad (1.10)$$

where the root-mean-square pressure p_{rms} is defined by $p_{rms} \equiv \sqrt{\overline{p}^2}$.

If the disturbance is a sinusoidal plane wave, then the mean square pressure, averaged over a period T, is:

$$\overline{p}^2 = \frac{p_0^2}{T} \int_0^T \sin^2 (kx - \omega t) dt$$

or:

$$\overline{p}^2 = \frac{p_0^2}{2T} \int_0^T [1 - \cos 2(kx - \omega t)] dt$$

The average of the cosine term over a period is zero. So that, for a sinusoidal plane wave:

$$\overline{p}^2 = p_0^2/2$$

and:

$$p_{rms} = p_0/\sqrt{2} = 0.707\, p_0 \qquad (1.11)$$

The intensity of a sinusoidal plane wave is:

$$I_{spw} = 0.325 \cdot 10^{-12} p_0^2 \qquad (1.12)$$

The quantity ρc has come to be called the specific acoustical resistance of the medium in accordance with an analogy set up with electrical circuits. According to this analogy:

$V \leftrightarrow F(= pA$; pressure times area)

$I \leftrightarrow u$

$R \leftrightarrow \rho c A$

$V = RI \leftrightarrow F = (\rho c A)u; p = \rho c u$

where ρc is the specific (per unit area) resistance. In general, the specific acoustical impedance is the ratio of the pressure to the velocity.

---

* For pressure measured in micropascals equation 1.10 becomes:

$$I = 0.65 \cdot 10^{-2} p_{rms}^2 \qquad (1.10a)$$

Thus far, ρc appears to be purely resistive, as it is for plane waves in an elastic medium. The important point concerning a purely resistive case is that the pressure and the velocity are in phase. Non-elasticity in the medium or, as we shall see shortly, the geometry of the field can cause a relative phase shift. In these cases the acoustical resistance becomes acoustical impedance and is a complex quantity.

The speed of sound in air is 34,400 cm/sec and in sea water it is about 150,000 cm/sec; the density of air is 0.00129 g/cm^3 and of sea water is 1.02 g/cm; the specific acoustical resistances are 42 and 154,000 acoustical ohms, respectively. Their ratio is:

$$\frac{(\rho c)_w}{(\rho c)_a} = 3700$$

The ratio may also be written:

$$\frac{(\rho c)_w}{(\rho c)_a} = \frac{u_a p_w}{u_w p_a} = \frac{p_w^2 I_a}{p_a^2 I_w}$$

For a given pressure disturbance, the particle speeds and displacements are about 4000 times greater in air than in water; for a given particle speed, the pressures in water are 4000 times greater than in air.

This means that the characteristics of a hydrophone or projector must be considerably different from those of a microphone or loudspeaker and that an air—water interface represents a strong discontinuity for the propagation of sound. Therefore, the interface is a good sound reflector.

The average power emitted by a sound source during the time τ (neglecting time delay due to finite propagation speeds) is obtained by integrating the intensity $I(\tau)$ over a surface Σ completely enclosing the source, viz.:

$$P(\tau) = \int_\Sigma I(\tau) \mathrm{d}s \tag{1.13}$$

As an example consider the power emitted by an isotropic point source at the origin of a spherical coordinate system. In this case intensity is a function of the radial coordinate (r) only. The power flowing through any sphere in radius r is:

$$P = \int_{-\pi/2}^{\pi/2} \int_0^{2\pi} I(r) r^2 \sin\theta \, \mathrm{d}\theta \, \mathrm{d}\phi$$

The integration results in:

$$P = 4\pi I r^2 .$$

Thus, we find that the intensity at any point in space due to an isotropic point source at $r = 0$ is:

$$I(r) = P/4\pi r^2 \tag{1.14}$$

This simple example serves to illustrate that power is the significant quantity to be related to the source while intensity is a field quantity. But, it is conventional to specify a standard measuring point and refer to "source intensity".

Source intensity is the intensity of the sound produced on the acoustic axis (direction of maximum sound propagation) one yard* from the acoustic center of the source (the ill-defined point from which rays emanate). To obtain source intensity empirically one must measure the intensity on the acoustic axis at a great distance and extrapolate, using equation 1.14, back to $r = 1$ yd.

For the case of an isotropic point source:

$$I_{\text{source}} = P/4\pi(91.5)^2$$

where P is in watts, I_{source} in watt/cm^2, and 91.5 cm is one yard. Then:

$$I_{\text{source}} = 0.95 \cdot 10^{-5} P \tag{1.15}$$

Our discussion concerning the intensity field in the vicinity of an isotropic point source has thus far been without reference to the pressure field. Let us now consider the pressure itself for this simple case.

In a spherically symmetric field the field variables are functions of the radial coordinate (and time) only. The wave equation reduces to:

$$\frac{1}{r^2}\frac{\partial}{\partial r}\left(r^2 \frac{\partial p}{\partial r}\right) = \frac{1}{c^2}\frac{\partial^2 p}{\partial t^2} \tag{1.16}$$

As may be verified by direct substitution:

$$p = \frac{a}{r} \exp\left[j(kr - \omega t)\right]$$

or:

$$p = \frac{a}{r} \cos(kr - \omega t) + j\,\frac{a}{r} \sin(kr - \omega t) \tag{1.17}$$

where a is a real constant related to the source strength and $j = \sqrt{-1}$, is a solution. The complex form of the mathematical solution has been taken in order that the manipulations may be performed with greater ease. One should be aware of the fact that the physical solution may be either the real or imaginary part.

A relation between the particle velocity and the pressure may be found by the application of equation 1.7 in the form:

$$u = \frac{1}{j\omega\rho} \operatorname{grad} p \tag{1.18}$$

---

* A yard is currently the most commonly used reference distance; however, the trend is toward the eventual use of a meter.

where the time integration has been performed. For our simple case this becomes:

$$u = \frac{1}{j\omega\rho}\frac{\partial p}{\partial r}$$

or, having performed the differentiation and returning the pressure to the equation:

$$u = \frac{1}{\rho c}\left(1 + \frac{j}{kr}\right)p^*$$

Or, we may solve it for pressure:

$$p = \frac{\rho c}{\left(1 + \dfrac{j}{kr}\right)} u = \frac{\rho c}{1 + \dfrac{1}{k^2 r^2}} \left(1 - \frac{j}{kr}\right) u \tag{1.19}$$

Hence we see that the specific acoustical impedance is complex and, therefore, the velocity and pressure have a non-zero relative phase.

If we assume that this spherical wave has locally plane wave fronts, the intensity is the average of the product of the real parts of p and u. The real pressure is:

$$\text{Re}\,(p) = \frac{a}{r}\cos\,(kr - \omega t)$$

and the real particle velocity is:

$$\text{Re}\,(u) = \frac{a}{\rho cr}\cos\,(kr - \omega t) + \frac{a}{\rho cr^2}\sin\,(kr - \omega t)$$

The second term of the velocity is 90° out of phase with the pressure and, like the reactive part of an electrical circuit, dissipates no power. As we have seen, the average of cosine squared is a half and the average of a sine—cosine product is zero. The intensity is, therefore:

$$I = \frac{1}{2\rho c}\left(\frac{a}{r}\right)^2 (10^{-7}) \tag{1.20}$$

Comparing this to equation 1.14 gives:

$$a = \sqrt{\frac{\rho c P}{2\pi} 10^{-7}} \tag{1.21}$$

It should be noted that the average intensity may also be obtained from:

---

* One often finds the imaginary part of this formula negative. It is positive here because of the sign adopted for the exponent in equation 1.17.

$$I = [\tfrac{1}{2} \mathrm{Re}\,(pu^*)] \cdot 10^{-7} \tag{1.22}$$

Attention should be brought to three other radiation field quantities: "Energy flux" is a vector having a magnitude that is equal to the power flowing across a unit surface and having a direction that is the direction of the outward normal to the surface. In the single beam case, the magnitude of the energy flux is $I \cos \theta$, where θ is the angle between the beam direction and the normal to the surface.

"Energy density" is the energy contained in a unit volume of the radiation field. It may be calculated from It/l, where l is the length of the segment of radiation that passes through the unit of area in time t, or more simply, the energy density is I/c.

The quantity $I\tau$ is sometimes used in the analysis of transient or non-stationary signals. It is the energy passing through the unit of area defining the intensity in time τ. Although this quantity is sometimes called "energy flux density", it has no physically significant name. The arbitrariness of τ is sometimes removed by specifying it to be one time unit (it then results in a unit like the "Langley" used in meterological radiology), or, in the case of transient signals, i.e., signals that die away fast enough for the time integral of the squared pressure to converge, τ is allowed to become infinite ($I\tau$ remains finite). In the latter case, it is the total energy impinging upon a unit area perpendicular to the beam.

1.3. LOGARITHMIC UNITS

It is quite common in acoustics to find energy related quantities measured in "decibels" relative to some reference. The decibel (dB) intensity is a logarithmic unit defined by:

$$L = 10 \log \frac{I}{I_{\mathrm{ref}}} \tag{1.23}$$

where L is meant to denote "level" and I_{ref} is some intensity reference value adopted by convention. An assortment of reference levels are found in the literature.

A rather simple, but little used, reference is 1 watt/cm^2. In this case:

$$L(\mathrm{re}\ 1\ \mathrm{W/cm^2}) = 10 \log I$$

The reason the use of this reference is not prevalent is that most levels that would result from its use would be negative. The conventional air acoustics reference level is the minimum intensity detectable by the average human ear at 1000 Hz which is 10^{-16} W/cm^2 (0.000204 dyn/cm^2 rms pressure). In this case:

$$L(\mathrm{re}\ 10^{-16}\ \mathrm{W/cm^2}) = 10 \log I + 160$$

The most common reference for underwater acoustics is the intensity (in water of a signal having an rms pressure of 1 dyn/cm^2 (μbar). This intensity is, according to equation 1.10:

$$I_{ref} = 0.65 \cdot 10^{-12}\,\mathrm{W/cm^2} \qquad (1.24)$$

so that the conventional intensity level for underwater acoustics is:

$$L(\mathrm{re}\ 0.65 \cdot 10^{-12}\,\mathrm{W/cm^2}) = 10 \log I + 122 \qquad (1.25)$$

where I is in watt/cm^2.

Although the reference level is the intensity of one μbar rms pressure, the reference is most often stated in terms of the pressure itself:

$$L(\mathrm{re}\ 1\mu\mathrm{bar}) = 10 \log I + 122 \qquad (1.26)$$

The level may be calculated in terms of pressure directly, i.e.:

$$L = 10 \log \frac{I}{I_{ref}} = 10 \log \frac{\bar{p}^2/\rho c}{(\bar{p}^2/\rho c)_{ref}}$$

The acoustic impedances will cancel provided the medium is not changed, so that:

$$L(\mathrm{re}\ 1\ \mu\mathrm{bar}) = 20 \log p_{rms} \qquad (1.27)$$

The difference between intensity levels may be expressed in dB without any reference level specified:

$$\Delta L = L_2 - L_1 = 10 \log \frac{I_2}{I_{ref}} - 10 \log \frac{I_1}{I_{ref}}$$

or:

$$\Delta L = 10 \log \frac{I_2}{I_1} \qquad (1.28)$$

or:

$$\frac{I_2}{I_1} = 10^{\Delta L/10}$$

These last formulae are quite useful in relating between ratios and decibels. It is handy to memorize at least the following conversions:

ΔL	0	3	6	7	9	10
I_2/I_1	1	2	4	5	8	10

Negative values of ΔL require that the ratio be inverted.

Another point to consider is that, quite often, there is more than just one discrete sinusoidal component in the sound wave, e.g., equation 1.4, or, of greater importance, there is a continuous set of components. In the latter case the pressure must be written as an integral of the continuous set of sinusoidal components over the frequency or wave number. But, more often intensity is of greater significance, so we will be concerned with so-called "power spectra" in greater detail.

Let $S_I(f)$ be the "power spectral density" defined by:

$$S_I(f) \equiv \lim_{\Delta f \to 0} \frac{\Delta I(f)}{\Delta f} \tag{1.29}$$

where $\Delta I(f)$ is the intensity of the signal in the frequency band Δf centered at frequency f.

In terms of an integral:

$$\Delta I(f) = \int_f^{f+\Delta f} S_I(f)\,\mathrm{d}f \tag{1.30}$$

Figure 1 illustrates possible distributions of spectral density over frequencies. Quite often we are satisfied with using these notions in an approximate sense. Let $W = \Delta f$, $I_w = \Delta I$, and we then approximate equation 1.29 by:

$$S_I(f) = \frac{I_w}{W} \tag{1.31}$$

This assumes that W is small.

These concepts are carried over to log units by the following definitions: Band Level (BL) is defined by:

$$BL = 10 \log \left(\frac{I_w}{I_{ref}}\right) \tag{1.32}$$

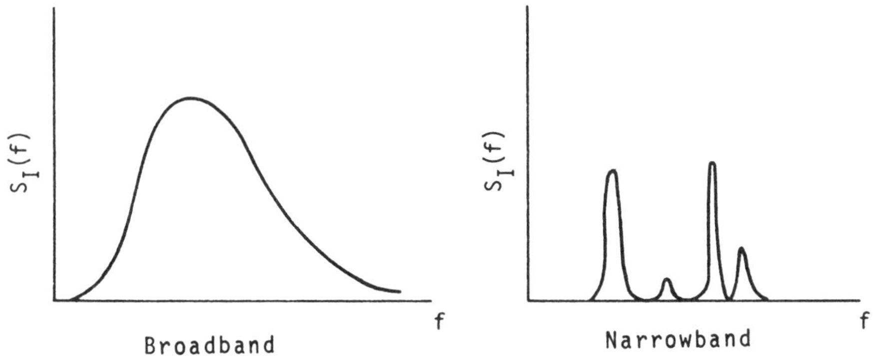

Fig. 1.

and Spectrum Level (*SPL*)* is defined by:

$$SPL = 10\log\left(\frac{S_I}{I_{\text{ref}}}\right) \tag{1.33}$$

In log units equation 1.31 becomes:

$$BL = SPL + 10\log W \tag{1.34}$$

PROBLEMS

1. What are the levels of the following intensities (re 1 μbar); 1.0, $0.65 \cdot 10^{-12}$, $1 \cdot 10^{-12}$, $2 \cdot 10^{-12}$, $5 \cdot 10^{-12}$, 10^{-11}, and 10^{-5} W/cm^2?
2. What are the levels of the following rms pressures; 1, 2, 10, 100, 10^{-2}, and 10^5 μbar?
3. What are the intensity ratios (I_2/I_1) of the following level differences $(\Delta L = L_2 - L_1)$; 0, 1, 2, 5, 10, 20, and -3 dB?
4. What are the spectrum levels for noise of -20 dB if measured in band widths of 1/2, 1, 2, 10, and 100 Hz?
5. Compare the relative pressures and particle motions for acoustic waves of equal intensity in air and in water.
6. Given a pressure at a point that is a superposition of two sinusoidal plane waves, i.e.:

   $$p(t) = p_1 \sin\omega_1 t + p_2 \sin\omega_2 t, \quad \omega_1 \neq \omega_2$$

   and assuming no dispersion, what is the average intensity of the wave? Give a general formula for the intensity at a point produced by the pressure:

   $$p(t) = \sum_{i=1}^{n} p_i \sin\omega_i t$$

7. What is the intensity of the exponentially damped sinusoidal plane wave:

   $$p(t) = p_0 \exp\left(-\frac{t}{\tau_0}\right)\cos\omega t$$

   if the averaging interval is $t = 0$ to $t = T$ (the period of the sinusoidal part)? Consider what happens when $\alpha = \tau_0/T$ goes to zero or infinity.
8. Show that the plane wave $p = f(kx - \omega t)$ is a solution of:

   $$\nabla^2 p - \frac{1}{c^2}\frac{\partial^2 p}{\partial t^2} = 0$$

---

* Do not confuse *SPL* used here and in Urick's *Principles of Underwater Sound for Engineers* with *SPL* used by some authors as "sound pressure level".

9. Show that for the plane wave $p = f(kx - \omega t)$ and $\boldsymbol{u} = u\,\boldsymbol{l}_x$, where $\boldsymbol{l}_x$ is a unit vector along the x-axis,

$$\text{grad}\, p = -\rho \frac{\partial \boldsymbol{u}}{\partial t}$$

results in $p = \rho c u$.

10. Verify that:

$$\overline{\text{Re}(a)\,\text{Re}(b)} = \tfrac{1}{2}\,\text{Re}(ab^*) = \tfrac{1}{2}\,a_0 b_0 \cos\phi$$

using $a = a_0 \mathrm{e}^{j\omega t}$ and $b = b_0 \mathrm{e}^{j(\omega t + \phi)}$, where a_0 and b_0 are real, and ϕ is the relative phase.

CHAPTER 2

ELECTROACOUSTIC AND CHEMICAL TRANSDUCTION

According to at least one definition of a transducer, it is a device that converts energy from one form to another.* In the case of an electroacoustic transducer, this conversion may be from electrical energy to acoustical energy (e.g., as a loudspeaker in air or a "projector" in water), or it may be from acoustical energy to electrical (e.g., as a microphone in air or a "receiver" or "hydrophone" in water). To include the one-way conversion of chemical energy into acoustical energy, we shall discuss explosive sources in this chapter on transduction.

2.1. ELECTROACOUSTIC TRANSDUCERS

Electroacoustic transducers designed for underwater application have the same purpose as the loudspeaker and the microphone used in air: they are to efficiently convert electrical to acoustical energy and vice versa. But the desire to have a maximum power conversion efficiency prohibits the direct adaptation of the loudspeaker to an underwater projector or the microphone to a hydrophone.

The reason that air transducers would not work efficiently in water is that they are designed to match the acoustical impedance of air, and, since the acoustical impedance of water is about 4000 times greater, the mismatch would be tremendous. The underwater transducer must operate with 60 times the force and 1/60 the displacement of a transducer in air. Actually, underwater transducer materials are not perfectly matched to the acoustical impedance of water but are 10 to 30 times larger. Although this mismatch is not small, it is better than would be obtained with loudspeakers and microphones used in water.

The most important single characteristic of a transducer material is the electromechanical coupling factor. This factor is related to the ratio of stored mechanical energy to total electrical input energy, and is related to the acoustical impedance match. It has also been shown that, for hydrophones used for detecting small signals, the self-noise of the receiver is more

* The term "transducer" is actually given a broader definition than this, but the above definition is exactly what we wish to use throughout this book.

important than the sensitivity (i.e., the voltage to pressure proportionality constant). In this chapter we shall consider these important quantities in greater detail.

In practice underwater electroacoustic transduction is accomplished by either of two phenomena*:

Electrostriction — the conversion of energy between acoustical and electrical forms by means of a dependence between electric fields and particle displacements in ferroelectric or piezoelectric materials.

Magnetostriction — the conversion of energy between acoustical and electrical forms by means of a dependence between magnetic fields and particle displacements in ferromagnetic materials.

The latter, although having certain special applications, is not in general usage and will not be discussed in these notes.

Although terminology applied to electroacoustic phenomena is not standardized, we will adhere to the following nomenclature: electrostriction is the general phenomenon as previously defined, and a distinction will be made between the piezoelectric effect (piezoelectric crystals) and the ferroelectric effect (ferroelectric ceramics).

Piezoelectricity

The unstrained state of a crystalline lattice represents an electrically neutral state. When certain lattices are strained, charge shifts causing excess charge to appear on the faces of the crystal, i.e., an electric field appears across the material. Conversely, if an electric field is applied the crystal will distort. Examples of piezoelectric materials are quartz, ammonium dihydrogen phosphate (ADP), Rochelle salt, and lithium sulphate.

A diagram explaining this phenomenon in quartz is given in Fig. 2.

For this form of electrostriction, the displacement and electrical field are proportional and therefore the acoustical and electrical frequencies are the same.

Advantages of piezoelectric crystals are uniformity in production and the linearity of the displacement—electric field relation. Among the disadvantages are its high electrical impedance (thus requiring high operating voltages which limit the power due to dielectric breakdown) and limitations as to the form in which it can be supplied.

Ferroelectricity

Ferroelectricity is an electrical phenomenon that is analogous to the ferromagnetic phenomenon but has nothing whatsoever to do with ferric

* Others exist but are not commonly used.

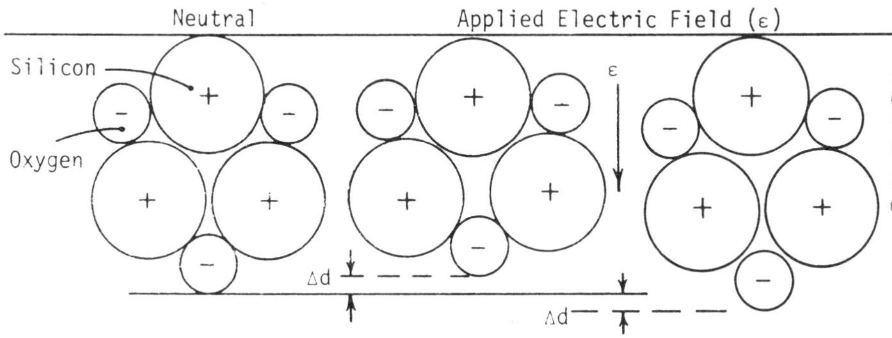

Fig. 2.

materials. A ferroelectric material is a material in which a permanent electric polarization can be established in an unstrained material just as a permanent magnetic field can be established in a ferromagnetic material. Electrostriction is accomplished by the reorientation of ferroelectric domains in an applied electric field.

Some such ferroelectric materials are ceramics of barium titanate, lead metaniobate, and lead zirconate-titanate. Some common trade names used in the literature are:

PZT (#) (Clevite brand lead zirconate-titanate),

Ceramic B (Clevite brand barium titanate), and

LM (Gulton G-2000 brand lead metaniobate).

A simplified diagram of the ferroelectric effect is shown in Fig. 3.

In ferroelectric materials the displacement is proportional to the square of the applied field. Therefore, in a natural state a ferroelectric material being driven by a sinusoidal electric field ($\epsilon_0 \sin \omega t$) would produce a sinusoidal displacement with twice the frequency, i.e.:

$$\Delta d \propto \sin^2 \omega t \;=\; \tfrac{1}{2}(1 - \cos 2\omega t)$$

To reduce this undesired effect, and to permit the use of the material as a receiver, a permanent electric polarization is established in the ceramic.

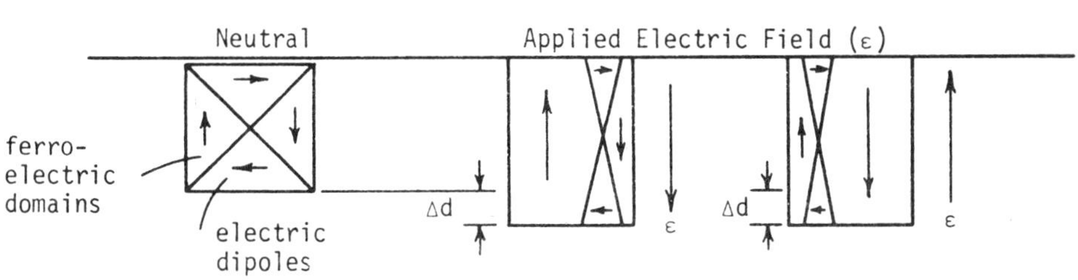

Fig. 3.

Thus, when the electric field ($\epsilon_0 \sin \omega t$) is applied it is in addition to an effective DC field (ϵ_{DC}). The following field results:

$$\epsilon^2 = (\epsilon_{DC} + \epsilon_0 \sin \omega t)^2 = \epsilon_{DC}^2 + 2\epsilon_{DC}\epsilon_0 \sin \omega t + \epsilon_0^2 \sin^2 \omega t$$

If $\epsilon_{DC} \gg \epsilon_0$, the last term is small and, since the DC field does not produce a variation in the displacement, the displacement may be written:

$$\Delta d \propto 2\epsilon_{DC}\epsilon_0 \sin \omega t$$

Note that there is no frequency doubling.

Ferroelectric ceramics have the disadvantages of being weak in tensile strength and variable in their properties. The former is generally what limits their power output, but prestressing can improve this situation. One of the major advantages of this material is that transducers can be made in any desired shape and size.

Electrical and mechanical systems

In order to provide a background in which the operation of a transducer may be more readily understood, we will review the fundamentals of electrical and mechanical oscillating systems. Let us begin with the electrical system.

The circuit in Fig. 4 represents a series RLC circuit.

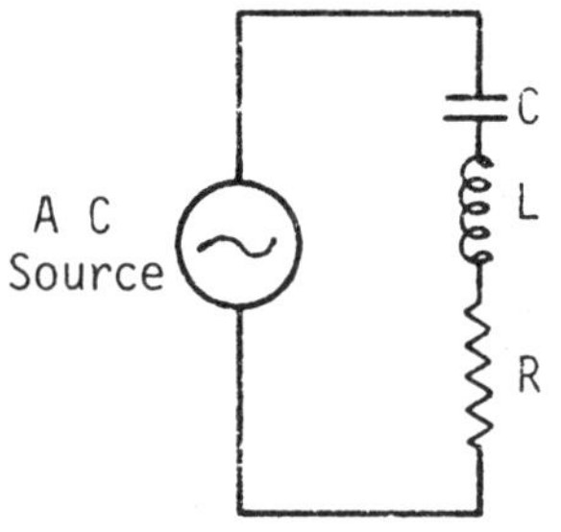

Fig. 4.

Suppose that the AC source is driving the system with the sinusoidal voltage $V = V_0 e^{j\omega t}$. The system satisfies:

$$L\ddot{q} + R\dot{q} + q/C = V_0 e^{j\omega t}$$

where q is electric charge and $\dot{q}$ is electric current ($I = \mathrm{d}q/\mathrm{d}t$). The steady-state solution is given by:

$$q = q_0 e^{j\omega t}$$

where:

$$q_0 = \frac{V_0}{-\omega^2 L + j\omega R + 1/C}$$

But we are more interested in the electric current which is given by:

$$I = j\omega q_0 e^{j\omega t}$$

where:

$$I = j\omega q_0 e^{j\omega t} = \frac{j\omega V_0 e^{j\omega t}}{j\omega R - \omega^2 L + 1/C}$$

This may be put into the form:

$$V = \left[R + j\left(\omega L - \frac{1}{\omega C}\right)\right] I$$

The coefficient of I is called the electrical impedance (Z):

$$Z = R + j\left(\omega L - \frac{1}{\omega C}\right) \tag{2.1}$$

with the real part (R) being called resistance and the imaginary part ($\omega L - 1/\omega C$) being called the reactance (X). Then:

$$Z = R + jX \tag{2.2}$$

We will also have occasion to use "admittance" (Y) which is defined by:

$$Y = 1/Z$$

Now we consider the mechanical system. Figure 5 represents a mass suspended from a rigid support by a spring in a viscous medium having a drag coefficient b.

If this system is driven by the force $F = F_0 e^{j\omega t}$, the equation for its motion is:

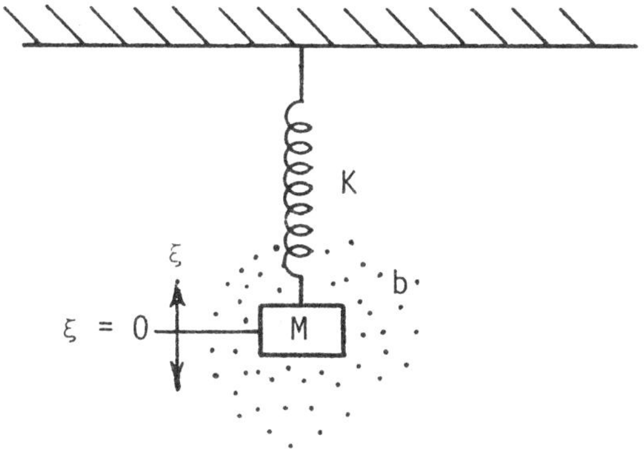

Fig. 5.

$$m\ddot{\xi} + b\dot{\xi} + K\xi = F_0 e^{j\omega t}$$

where m is the mass and K the spring constant. The steady-state solution is given by:

$$\xi = \xi_0 e^{j\omega t}$$

By substitution into the differential equation:

$$\xi_0 = \frac{F_0}{-\omega^2 m + j\omega b + K}$$

We are particularly interested in the velocity $u = \mathrm{d}\xi/\mathrm{d}t$. It is:

$$u = j\omega\xi_0 e^{j\omega t} = \frac{j\omega F_0 e^{j\omega t}}{j\omega b - \omega^2 m + K}$$

Dividing through by $j\omega$ and solving for F, we get:

$$F = \left[b + j\left(\omega m - \frac{K}{\omega}\right)\right] u \tag{2.3}$$

In a manner analogous to the electrical case, we call the coefficient of u in (2.3) the mechanical impedance (Z_m), the real part (b) mechanical resistance (R_m), and the imaginary part ($\omega m - K/\omega$) mechanical reactance (X_m). Then:

$$Z_m = b + j\left(\omega m - \frac{K}{\omega}\right) = R_m + jX_m \tag{2.4}$$

It is readily shown that the condition of resonance is given by X or X_m equal to zero in the electrical or mechanical systems, respectively. The formulations serve to illustrate a close analogy between electrical and mechanical systems and should allow for an easy understanding of the acoustic transducer.

Radiated energy

Thus far we have considered mechanical systems that are not radiating energy. We now let the mechanical system by a vibrating transducer radiating acoustic energy into the water. This will cause an additional impedance which we will call the radiation impedance (Z_r). In general, it has a resistive part (R_r) which is in phase with the force and represents energy actually radiated into the water, and a reactive part (X_r) which is 90° out of phase with the force and produces no radiated energy.

If we consider a transducer with a plane surface of area A, having dimensions large compared to the wavelength of the sound produced, then we may write:

$$p = \rho c u \quad \text{or} \quad F = \rho c A u \tag{2.5}$$

where F and u are the instantaneous force on the fluid at the transducer face and the velocity of the face, respectively. And ρc is the specific acoustic impedance of the water.

In this case the impedance is real with resistance:

$$R_r = \rho c A$$

and no reactance (i.e., $X_r = 0$). The reactance becomes non-zero as the edge effects become important.

In general, we may write the radiation impedance as:

$$Z_r = \rho c A(r + jx) \tag{2.6}$$

where $r \to 1$ and $x \to 0$ as the acoustic wave becomes planar. The solid curves in Fig. 6 illustrate the variation of r and x for a disc projector as a function of ka, where k is the angular wave number and a is the radius of the disc.

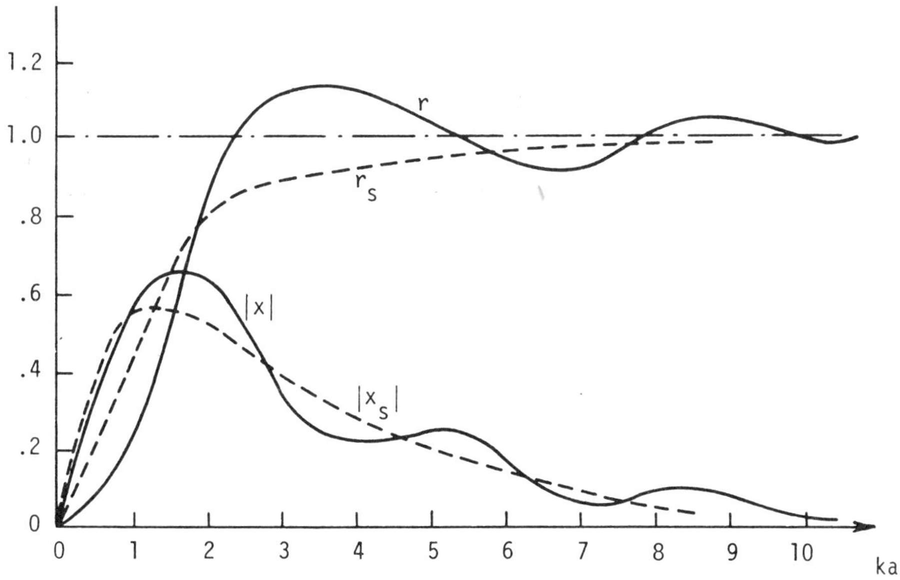

Fig. 6.

The specific acoustic impedance of a spherical wave (as we found in Chapter 1) is:

$$\rho c \left(\frac{k^2 r^2}{1 + k^2 r^2} - \frac{jkr}{1 + k^2 r^2} \right)$$

Thus the radiation impedance of a pulsating sphere of radius a is:

$$Z_r = \rho c A \left(\frac{k^2 a^2}{1 + k^2 a^2} - \frac{jka}{1 + k^2 a^2} \right)$$

Plotted in the diagram of Fig. 6 as dashed lines are:

$$r_s = \frac{k^2a^2}{1+k^2a^2}$$

$$|x_s| = \frac{ka}{1+k^2a^2}$$

Generalized theory of equivalent circuits

Rather than discussing specific transducers and their various parameters, we will present the generalized relationships between the electrical and mechanical elements of the transducer. The relations are sometimes called the "4-pole equations" because, according to this analysis, the transducer is represented as a four-terminal network with an electrical and a mechanical input as shown in Fig. 7.

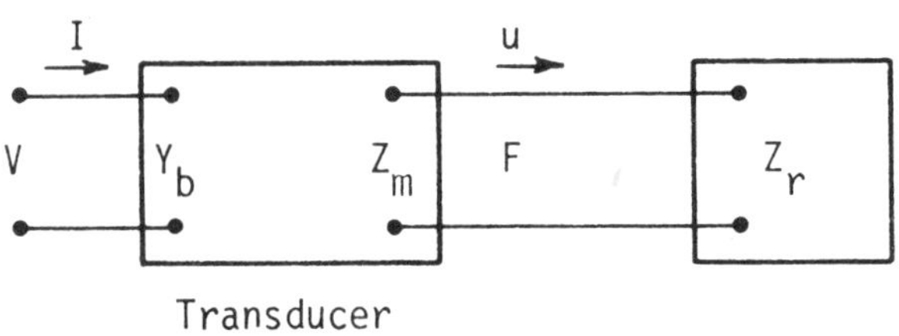

Fig. 7.

The current (I) flowing into the circuit is related to the voltage (V) across the terminals and the velocity (u) of the transducer face according to:

$$I = Y_b V - \phi u \qquad (2.7)$$

Y_b is the blocked (or clamped) input admittance corresponding to $u = 0$, and ϕ is the transformer ratio relating the short-circuit current ($V = 0$) to the velocity of the transducer face.

From another point of view, we find that the force (F) acting on the face of a transducer is related to the velocity of the face and the voltage produced at the electrical side according to:

$$F = \phi V + Z_m u \qquad (2.8)$$

where Z_m is the mechanical impedance defined earlier; we now call it the short-circuited mechanical impedance since, in the coupled system, it is determined when $V = 0$.

Proceeding to the next branch of the circuit, we find that the force back on the transducer face may be calculated from:

$$F = -Z_r u \qquad (2.9)$$

The negative sign appears because this is the reaction force to the output force of the transducer. Using this to eliminate F from equation 2.8, then substituting for u in (2.7) we get:

$$I = \left(Y_b + \frac{\phi^2}{Z_m + Z_r}\right) V$$

Then the input admittance is:

$$Y_i = Y_b + Y_m \tag{2.10}$$

where Y_m is the motional admittance given by:

$$Y_m = \frac{\phi^2}{Z_m + Z_r} \tag{2.11}$$

For an electrostrictive transducer, the blocked admittance is simply the admittance of a parallel resistance (R_0) and capacitance (C_0) and is therefore:

$$Y_b = \frac{1}{R_0} + j\omega C_0 \tag{2.12}$$

and the mechanical impedance is given by:

$$Z_m = b + j\left(\omega m - \frac{K}{\omega}\right) = R_m + jX_m$$

as we saw earlier. Here m represents the effective mass of the transducer and K the effective elastic constant.

Figure 8 shows the equivalent circuit for a transducer emitting plane waves.

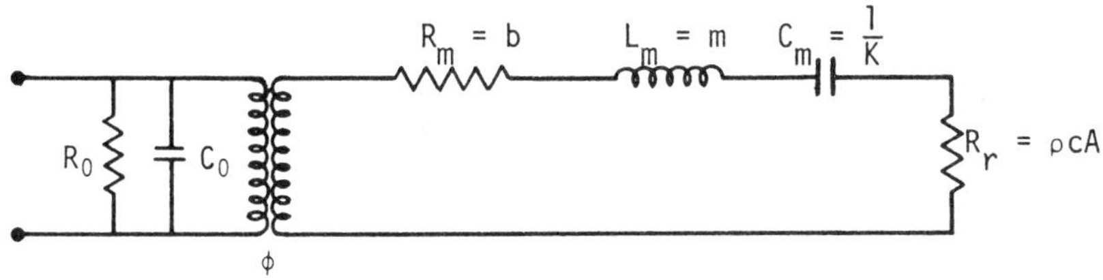

Fig. 8.

If the following quantities are defined:

$$R_{\mathrm{R}} = \frac{R_{\mathrm{r}}}{\phi^2}; \quad R_{\mathrm{M}} = \frac{R_{\mathrm{m}}}{\phi^2}; \quad L_{\mathrm{M}} = \frac{L_{\mathrm{m}}}{\phi^2}; \quad C_{\mathrm{M}} = \phi^2 C_{\mathrm{m}} \tag{2.13}$$

equation 2.10 may be written:

$$Y_i = \frac{1}{R_0} + j\omega C_0 + \left[R_{\mathrm{R}} + R_{\mathrm{M}} + j\left(\omega L_{\mathrm{M}} - \frac{1}{\omega C_{\mathrm{M}}}\right)\right]^{-1} \tag{2.14}$$

Figure 9 illustrates the equivalent circuit in terms of these parameters.

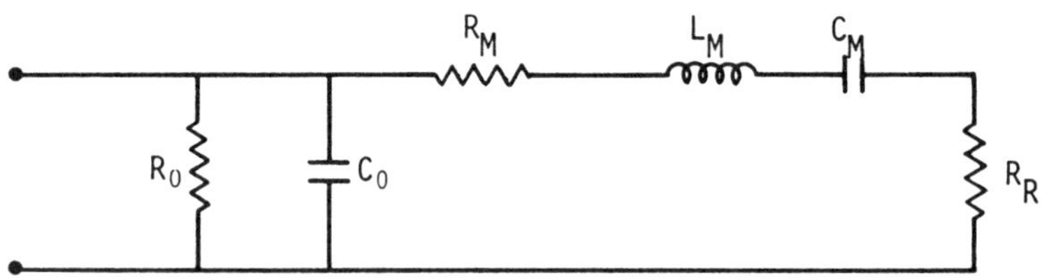

Fig. 9.

Power is dissipated into the three resistive elements of the circuit. The power losses in R_0 and R_{M} are due to the joule heating and mechanical losses, respectively. The power dissipated into R_{R} is the radiated acoustic power. Power in the reactive elements is stored and not available for acoustic radiation.

Ideally:

$$R_0 \to \infty, \quad R_{\mathrm{M}} \to 0, \quad \text{and} \quad \omega L_{\mathrm{M}} - \frac{1}{\omega C_{\mathrm{M}}} = 0$$

In practice it is possible to obtain a zero in the reactance, but this occurs only at resonant frequency (ω_{res}). Sometimes it is desirable to have a transducer operate at a single frequency. In this case a transducer with a sharp resonance at that frequency is sought. But often transmitters and receivers are expected to operate over a broad band, so transduction is oftentimes sought well away from these resonances.

Power conversion and impedance matching

Just as intensity is given by equation 1.22, the average power in an electrical element may be expressed as:

$$P = \tfrac{1}{2}\mathrm{Re}[VI^*]$$

For sinusoidal voltage across the element and current through the element with amplitudes V_0 and I_0, respectively, and relative phase ϕ*, this equation may be reduced to:

$$P = \tfrac{1}{2} V_0 I_0 \cos\phi = V_{rms} I_{rms} \cos\phi$$

where $\cos\phi$ is the power factor. In order to achieve maximum power conversion in the element for given voltage and current amplitudes, it is desirable to have the power factor equal to unity (i.e., $\phi = 0$).

Consider Fig. 10.

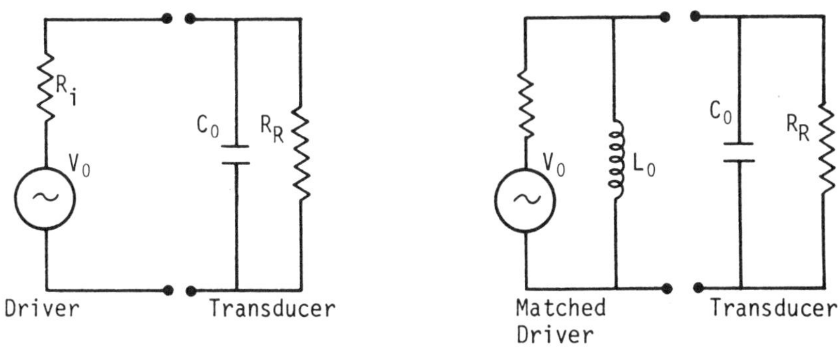

Fig. 10.

In the first circuit the transducer is being driven at its resonance frequency by a source with voltage amplitude V_0 and internal resistance R_i. The power factor is given by:

$$\cos\phi = (1 + \omega_{res}^2 C_0^2 R_R^2)^{-1/2}$$

In the second circuit an inductance (L_0) is added to tune out the capacitive reactance. If the magnitude of the inductive reactance is equal to the magnitude of the capacitive reactance (i.e., $\omega L_0 = 1/\omega C_0$) the power factor is unity.

The optimum operating condition is obtained when the source reactance is equal and opposite the transducer reactance and the source resistance is equal to the transducer resistance. This is, the optimum operating condition occurs when:

$$Z_{source} = Z^*_{transducer}$$

or:

$$R_s + jX_s = R_t - jX_t$$

* Not the ϕ of the previous section.

This conjugate matched impedance is obtained using a matching transformer, which provides the proper source resistance and inductive reactance. The circuit is shown in Fig. 11.

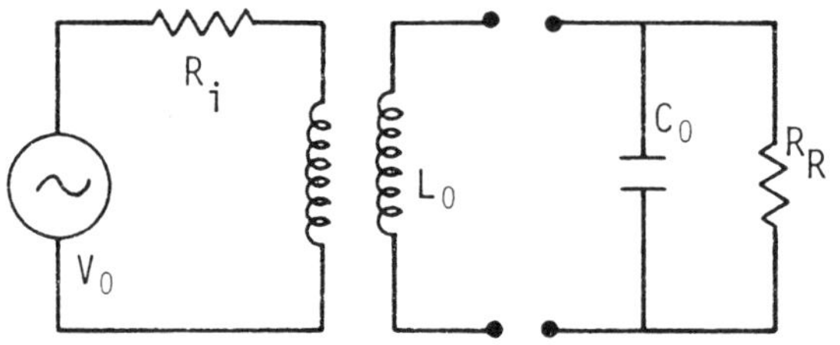

Fig. 11.

Transmission efficiency

The efficiency of a transducer is the ratio of the power radiated to the total input electrical power. To simplify the calculations we assume that the electrical capacitance is tuned out by the source and calculate the efficiency at resonance so that all the reactance is eliminated from the circuit. The equivalent circuit is illustrated in Fig. 12.

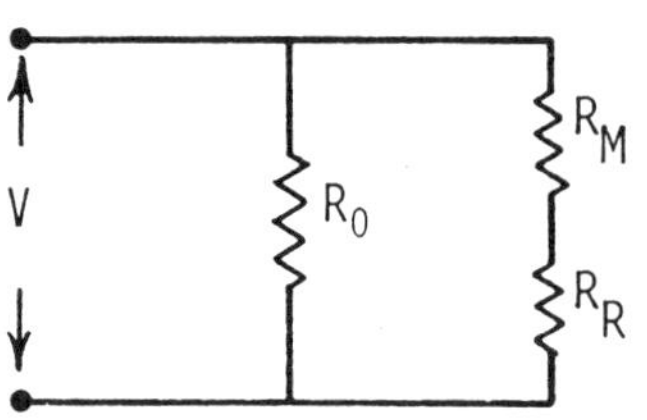

Fig. 12.

The total power dissipated in the transducer is:

$$P_t = \frac{V_{\mathrm{rms}}^2}{\left(\dfrac{1}{R_0} + \dfrac{1}{R_M + R_R}\right)^{-1}} = \frac{V_{\mathrm{rms}}^2 (R_M + R_R + R_0)}{R_0(R_M + R_R)}$$

The power radiated is:

$$P_R = I_{\mathrm{rms}}^2 R_R$$

where I_{rms} is the rms current in the motional branch. Then:

$$P_R = \left(\frac{V_{\mathrm{rms}}}{R_M + R_R}\right)^2, \ R_R = \frac{V_{\mathrm{rms}}^2 R_R}{(R_M + R_R)^2}$$

The efficiency is:

$$\eta = \frac{P_R}{P_t} = \frac{R_0 R_R}{(R_M + R_R)(R_M + R_R + R_0)}$$

This is the maximum internal efficiency that may be obtained from a given transducer. The circuit external to the transducer also dissipates power. In particular, when the source is conjugate matched to the transducer 50% of the power is dissipated in the source.

Quality factor

A measure of the width of the resonant peaks is the quality factor (Q) of a circuit. It is defined by:

$$Q = \left(\frac{\text{Energy stored in the reactance}}{\text{Energy dissipated in the resistance}}\right)_{\text{per cycle}}$$

and is related to bandwidth by:

$$Q = \omega_{res}/\Delta\omega$$

where ω_{res} is the mechanical resonant frequency and $\Delta\omega$ is the bandwidth between half-power (3 dB) points.

The mechanical Q is given by:

$$Q_M = \frac{\omega_{res} L_M}{R_R + R_M} \tag{2.15}$$

R_M is generally small for crystal oscillators, but R_R depends upon the medium in which the crystal oscillates. R_R is zero for a crystal operating in a vacuum, so Q is very high ($\sim 10^6$). The same crystal acting in air would have a Q of about 10^4. These high Q-values are the reason that crystal oscillators produce very sharp tones. R_R is very large for a transducer operating in water and the Q is relatively small (~ 10); therefore, the resonance is broad.

Q_M determines the bandwidth of a projector if the source has a high output impedance. But, as we have seen, for optimum transduction, the source is conjugate matched to the transducer in which case the controlling Q is given by:

$$Q = \frac{Q_e + Q_M}{2}$$

where Q_e is the quality factor of the electrical branch. The electrical quality factor may be calculated from:

$$Q_e = \omega_{res} C_0 (R_R + R_M)$$

Electromechanical coupling factor

At low frequencies the electrostrictive transducer behaves essentially like two capacitors in parallel: the electrical capacitance (C_0) and the mechanical capacitance (C_M) produced by the elasticity of the transducer. When the voltage V is applied to the parallel circuit, electrical energy equal to $C_0 V^2/2$ is stored in C_0, and mechanical energy $C_M V^2/2$ is stored in C_M. The ratio of mechanical to electrical energy is given by:

$$\frac{C_M V^2/2}{C_0 V^2/2} = \frac{C_M}{C_0}$$

For efficient transduction, C_M/C_0 should be large. The electromechanical coupling factor k is a measure of this efficiency. It is defined as:

$$k^2 = \frac{\dfrac{\pi^2}{8}\dfrac{C_M}{C_0}}{1 + \dfrac{\pi^2}{8}\dfrac{C_M}{C_0}} \tag{2.16}$$

This is often approximated by:

$$k = \sqrt{\frac{\pi^2 C_M}{8 C_0}} \tag{2.17}$$

Although we do not wish to go into specific transducer materials and geometries, it should be pointed out that k can be related to parameters of the material and is constant for a given crystal cut and mode of vibration. The value of k is 0.1 for quartz and 0.18 for barium titanate. In the next section, we shall see how the electromechanical coupling factor is related to receiver sensitivity of a hydrophone.

Receiver sensitivity

Equations 2.7 and 2.8 may be used to determine a relation for the open circuit receiver sensitivity. In this case $I = 0$ and equation 2.7 becomes:

$$u = \frac{Y_b}{\phi} V$$

This is substituted into equation 2.8 to give:

$$F = \left(\phi + \frac{Z_m Y_b}{\phi}\right) V = \phi \left(1 + \frac{Z_m Y_b}{\phi^2}\right) V$$

Let us define the receiver sensitivity by:

$$\alpha = \frac{|V|}{|p|} \tag{2.18}$$

or in terms of force:

$$\alpha = \frac{A|V|}{|F|}$$

Then, from the force—voltage relation above:

$$\alpha = \frac{A}{\phi}\left|\frac{1}{1 + Z_M Y_b}\right|$$

where Z_M has replaced Z_m/ϕ^2. Replacing the blocked admittance by the blocked impedance, we find:

$$\alpha = \frac{A}{\phi}\left(\frac{|Z_b|}{|Z_b + Z_M|}\right) \tag{2.19}$$

Or, writing out the impedances:

$$\alpha = \frac{A}{\phi}\left[\frac{\left|\left(\frac{1}{R} + j\omega C_0\right)^{-1}\right|}{\left|\left(\frac{1}{R_0} + j\omega C_0\right)^{-1} + R_R + R_M + j\left(\omega L_M - \frac{1}{\omega C_M}\right)\right|}\right]$$

If we assume that R_0 is infinite and that R_M and R_R are zero (the latter since no energy is radiated), we simplify this to:

$$\alpha = \frac{A}{\phi}\left[\frac{\left|\frac{1}{j\omega C_0}\right|}{\left|\frac{1}{j\omega C_0} + j\left(\omega L_M - \frac{1}{\omega C_M}\right)\right|}\right]$$

or:

$$\alpha = \frac{A}{\phi}\left(\frac{1/\omega C_0}{\left|\omega L_M - \frac{1}{\omega C_M} - \frac{1}{\omega C_0}\right|}\right)$$

finally:

$$\alpha = \frac{A}{\phi}\left[\frac{C_M/C_0}{\left|(\omega^2 L_M C_M - 1) - \frac{C_M}{C_0}\right|}\right] \tag{2.20}$$

At resonance (i.e., $\omega_{res} = 1/\sqrt{L_M C_M}$):

$$\alpha = A/\phi$$

Below resonance (i.e., $\omega \ll \omega_{res}$):

$$\alpha = \frac{A}{\phi}\left(\frac{C_M/C_0}{\frac{C_M}{C_0}+1}\right) \sim \frac{A}{\phi}\frac{C_M}{C_0} = \frac{8}{\pi^2}\frac{A}{\phi}k^2$$

Above resonance (i.e., $\omega \gg \omega_{res}$):

$$\alpha = \frac{A}{\phi}\frac{1}{\omega^2 L_M C_0}$$

Figure 13 illustrates the sensitivity as a function of frequency.

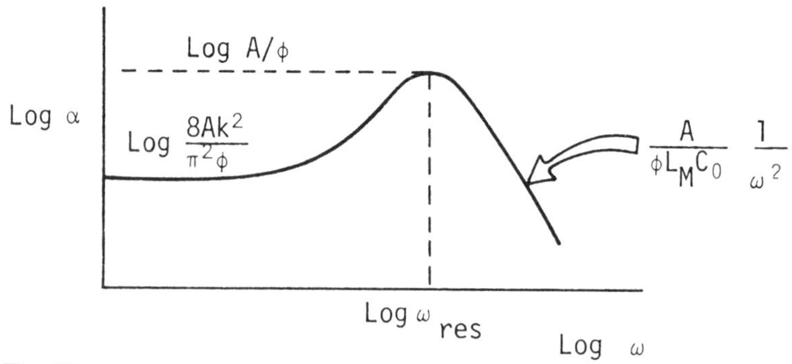

Fig. 13.

2.2. EXPLOSIVE SOURCES

Small charges of explosive materials, particularly TNT, are also used for sound sources. These have many applications ranging from seismic studies to anti-submarine warfare (ASW). The amount of charge used ranges from fractions of pounds to a few hundred pounds in weight. Their major advantage is the high source levels that they can produce, but a disadvantage is that they do not produce signals that lend themselves to very sophisticated signal processing.

The chemical energy of the explosion is converted into acoustical energy through the production of a rapidly expanding bubble of incandescent gas. As the bubble expands a shock wave of acoustic energy is radiated into the water.

Due to the inertia of the water, the bubble will overshoot its equilibrium position then eventually collapse producing a slightly negative pressure in the water. This cycle may occur several times. The additional pulses produced by this cycling are called bubble pulses.

Excluding the bubble pulses, the radiated acoustic pressure at a given point is:

$$p(t) = p_0 \exp(-t/t_0) \tag{2.21}$$

where time is measured from the time the leading edge of the shock wave reaches the point of interest, and t_0 is the time it takes for the pressure to reduce to p_0/e. The characteristic time and the peak pressure are functions of both the charge weight and the range.

Empirically it has been found* that TNT explosions may be described by:

$$p_0 = 2.16 \cdot 10^4 \left(\frac{W^{1/3}}{r}\right)^{1.13}$$

$$t_0 = 0.058\, W^{1/3}\left(\frac{W^{1/3}}{r}\right)^{-0.22} \tag{2.22}$$

where r is in feet, t in milliseconds, W in pounds and p in lbs/in^2.

Due to attenuation of the high-frequency components that make up the pulse, the onset of the pulse will not always remain abrupt. Attenuation and non-linearity associated with the propagation of a high-intensity pulse will cause the pulse to broaden (through the factor $(W^{1/3}/r)^{-0.22}$ in t_0) and produce a spreading loss in excess of spherical (through the extra factor $(W^{1/3}/r)^{0.13}$ in p_0).

The total energy of the explosion and the volume of the charge are both proportional to the weight W. As a result of the latter proportionality, an appropriate scaling factor for linear measures is $W^{1/3}$. Thus we find that if we define a reduced time (t_r) and a reduced range (r_r) by:

$$t_r = tW^{-1/3} \quad \text{and} \quad r_r = rW^{-1/3}$$

respectively, the charge weight will be eliminated from the equations, i.e.:

$$p_0 = 2.16 \cdot 10^4 \left(\frac{1}{r_r}\right)^{1.13}$$

$$t_{r0} = 0.058\, r_r^{0.22} \tag{2.23}$$

The "energy flux density" $e(t)$ is defined to be:

$$e(t) = It = \frac{1}{\rho c}\int_0^t p^2(t')\mathrm{d}t'$$

So, for the initial pulse of a TNT explosion in water:

$$e(t) = \frac{p_0^2}{\rho c}\int_0^t \exp\left(-\frac{2t}{t_0}\right)\mathrm{d}t$$

$$e(t) = \frac{p_0^2 t_0}{2\rho c}\left[1-\exp\left(-\frac{2t}{t_0}\right)\right]$$

---

* Arons, A.D., Yennie, D.R. and Cotter, T.P., *Long Range Shock Propagation in Underwater Explosion Phenomena II*, U.S. Navy Dept. Bur. Ordnance NAVORD Rept. 478 (1949).

And the total energy flux density is given when $t \to \infty$. It is:

$$E = \lim_{t \to \infty} e(t) = \frac{p_0^2 t_0}{2\rho c}$$

The pressure impulse (Im) is defined by:

$$Im = \int_0^\infty p\mathrm{d}t$$

The impulse of the initial pulse is:

$$Im_0 = p_0 \int_0^\infty \exp(t/t_0)\mathrm{d}t = p_0 t_0$$

Using the equations of (2.22), the impulse of the initial pulse is:

$$Im_0 = 1.78\, W^{1/3} \left(\frac{W^{1/3}}{r}\right)^{0.94} \frac{\mathrm{lb.sec}}{\mathrm{in}^2}$$

The bubble pulses have the form of an exponential rise in pressure followed by a symmetric exponential decay. Each pressure peak in succeeding bubble pulses reduces by a factor of about 1/5. The time interval between the initial pulse and the first bubble pulse is given by:

$$T_1 = \frac{4.36\, W^{1/3}}{(d + 33)^{5/6}}$$

where d is the depth in feet of the detonation below the surface.

Figure 14 shows the pressure—time curve for the most significant portions of the explosion, Fig. 15 shows an impulse representation of it.

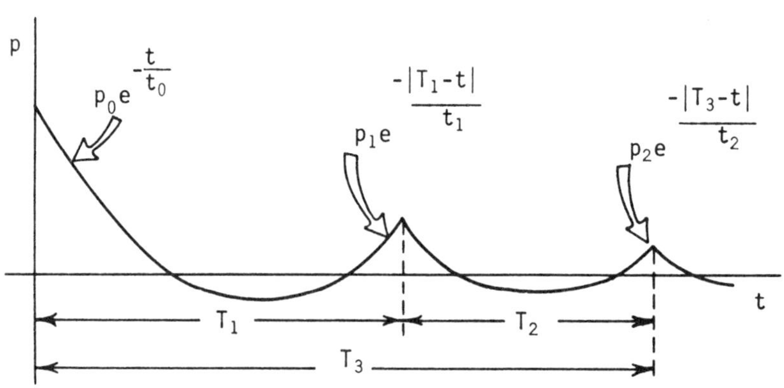

Fig. 14.

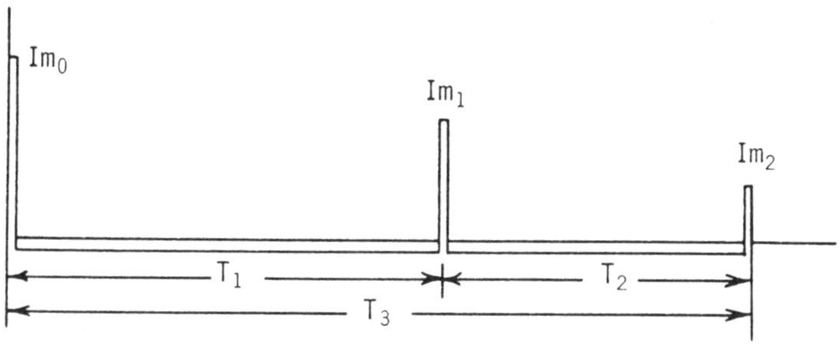

Fig. 15.

The latter figure is useful for calculating the low-frequency spectrum of explosion.

The amplitude spectrum is obtained from:

$$A(f) = \int_0^\infty p(t)\, e^{j\omega t} dt$$

and the energy flux density spectrum from:

$$E(f) = \frac{2AA^*}{\rho c}$$

The factor of two is introduced because we consider positive frequencies only.

Table I gives the pressure and impulse signature and their energy flux density spectra.

TABLE I

Description	Schematic shape	Spectrum equation
Shock	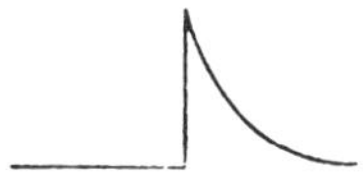	$E_0(f) = \dfrac{2p_0^2}{\rho c(1/t_0^2 + 4\pi^2 f^2)}$
Bubble		$E_1(f) = \dfrac{8}{\rho c}\left(\dfrac{p_1/t_1}{1/t_1^2 + 4\pi^2 f^2}\right)^2$
Impulse	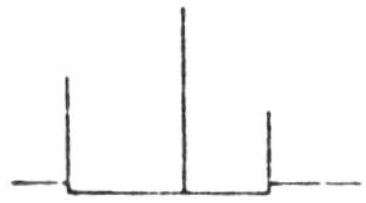	$E_I(f) = \left(\dfrac{2}{\rho c}\right)\left[\{I_0 + I_1\cos 2\pi f T_1 + I_2 \cos 2\pi f T_3 - N\sin 2\pi f T_3\}^2 + \{I_1 \sin 2\pi f T_1 + I_2 \sin 2\pi f T_3 - N(1 - \cos 2\pi f T_3)\}\right]^2$ where $N = I_0 + I_1 + I_2/2\pi f T_3$

From Weston, D.E., 1961. *Explosive Sources*, Ed. V.M. Albers, Inst. on Underwater Acoustics, Plenum Press, N.Y. Used with permission of Plenum Press.

The following graph* (Fig. 16) shows the combined form of the spectra.

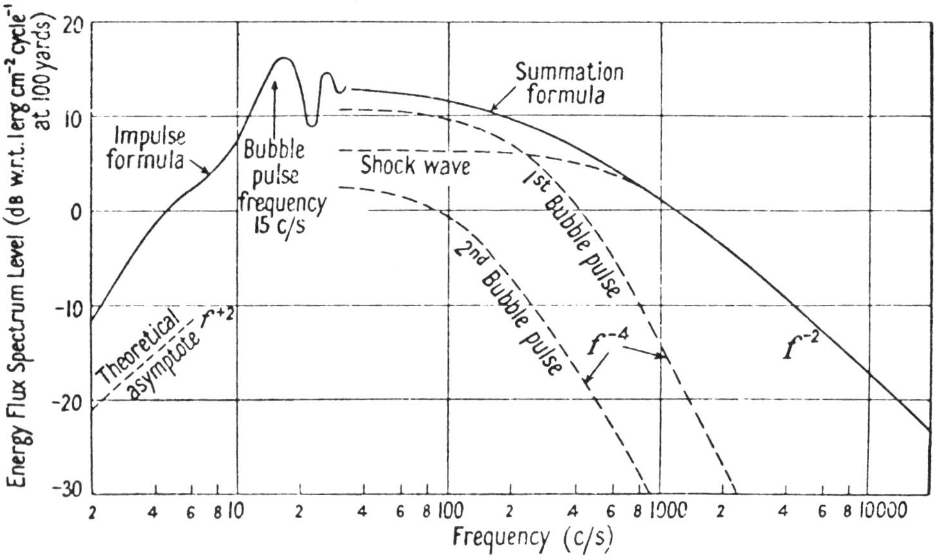

Fig. 16.

PROBLEMS

1. For the equivalent circuit (Fig. 17) calculate: (a) the resonance frequency; (b) the mechanical Q; (c) the electrical Q; (d) the bandwidth if the driver impedance is high; (e) the bandwidth if the driver is matched; (f) the efficiency; and (g) the electromechanical coupling coefficient.

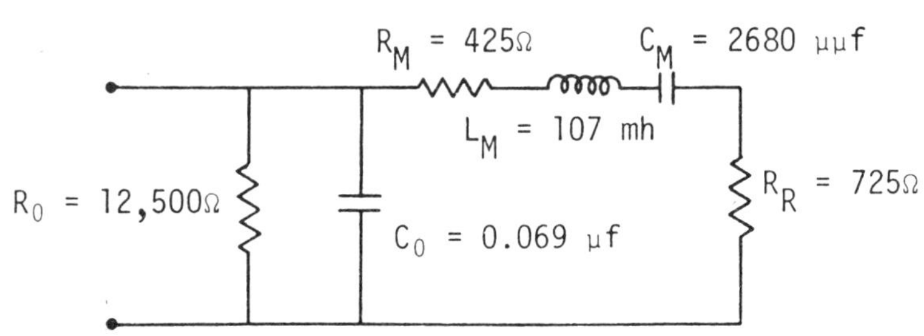

Fig. 17.

2. For the above transducer, plot the motional admittance in the complex plane in the range $0 \leqslant \omega < \infty$.

* From Weston, D.E., 1961. *Explosive Sources,* Ed. V.M. Albers, Inst. on Underwater Acoustics, Plenum Press, N.Y. Used with permission of Plenum Press.

3. Using 1/2 and 2 lb charge weights of TNT, calculate p_0, t_0, and total energy flux density at 100 yards. Calculate the time interval between the initial shock and the first bubble for both explosions occurring at 100, 1000, and 10,000 feet.
4. Show that for an explosive source $r^2 p_0^2 t_0$ should be independent of r; is it and if not why? Also, how does it depend upon W and is this dependence reasonable?

CHAPTER 3

HYDROPHONES, PROJECTORS, AND CALIBRATION

Since quantitative data are required in the study of underwater acoustics, it is necessary to define a receiving response for the electroacoustic devices used in measuring the sound field and a transmitting response for electroacoustic devices used in producing a sound field. It is also necessary that these devices be calibrated so that these responses may be accurately known.

3.1. TRANSDUCER RESPONSES

The receiver response (pressure sensitivity) of a hydrophone is the magnitude of the open-circuit voltage per unit magnitude of plane wave pressure incident on the hydrophone, i.e.:

$$\alpha = |V|/|p| \tag{3.1}$$

Or, in other words, the open-circuit rms voltage produced by a plane wave of unit rms pressure, i.e.:

$$\alpha = V_{rms}/P_{rms} \tag{3.1a}$$

Often the response is expressed in decibels:

$$\text{Receiver response} = 20 \log (\alpha/\alpha_{ref}) \tag{3.2}$$

The references response is taken to be one volt rms voltage produced by one dyn/cm^2 rms plane wave pressure, i.e.:

$$\alpha_{ref} = 1 \frac{\text{volt}}{\text{dyn/cm}^2} = 1 \frac{\text{volt}}{\mu\text{bar}}$$

So the response in dB (re 1 volt/μbar) is 20 log α.

The transmitter response of a projector is the magnitude of the pressure produced at a point one meter (or sometimes one yard) from the acoustic center in a direction along the acoustic axis per unit magnitude of electrical current in the projector, i.e.:

$$\beta = |p|/|i| \tag{3.3}$$

Or, the transmitter response may be defined in terms of rms values. The response is also expressed in decibels:

$$\text{Transmitter response} = 20 \log (\beta/\beta_{ref}) \qquad (3.4)$$

β_{ref} is taken to be 1 μbar/ampere, so the response in dB (re 1 μbar (at one meter)/ampere) is $20 \log \beta$.

Sometimes the transmitter response is referenced by 1 μbar (at one yard)/ampere. This response is 0.87 dB larger than the previous one because:

$$20 \log (1.1) = 0.87$$

where 1.1 is the number of yards per meter.

3.2 CALIBRATION

For most transducer applications, it is necessary that the responses be known. There are many ways to calibrate transducers*, but let us consider only the two simplest and most common: they being comparison and reciprocity methods.

Comparison method

The simplest and most direct way to calibrate a hydrophone or projector is to compare it with a standard. In the case of hydrophone calibration, a sound field is produced in an appropriate calibration tank and the responses of a standard and the unknown hydrophones are compared. In the case of projector calibration, the sound fields produced by the standard and the unknown are compared.

In spite of the fact that it is necessary to have standards available in order to use this method, it is the routine way transducers are calibrated. Such standards are available on loan or rental basis from naval agencies.

Reciprocity method

The reciprocity calibration method is more complicated than the comparison method but requires no standard. It is based upon the reciprocity theorem:

> If a generalized force whose magnitude $|\mathfrak{F}|$ is applied in any branch A of a system composed of linear elements and a response $|\mathfrak{I}|$ is measured in branch B, their ratio, called the transfer impedance, will be unchanged upon an exchange of the points of application.

Let branch A represent the electrical branch for which the generalized

---

* For a list of the various methods see table prepared by T.F. Johnston, appearing on pages 32 and 33 of Urick, R.M., 1967. *Principles of Underwater Sound for Engineers*, McGraw-Hill, N.Y.

force is the driving current (i) for the transducer acting as a projector, and the response is the open-circuit voltage (V) produced by the transducer acting as a hydrophone. Let branch B represent some point in the acoustic medium for which the pressure (p_B) at that point is the medium's response produced by the projector driven by the current (i), and the source strength (Q)* at the point is the generalized force that produces the open-circuit voltage response (V) in the transducer. Figure 18 illustrates these points.

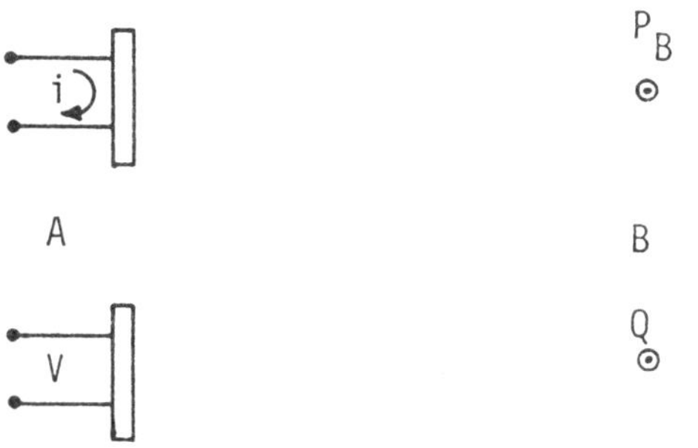

The reciprocity theorem states that:

$$\frac{|i|}{|p_B|} = \frac{|Q|}{|V|} \cdot 10^{-7} \tag{3.5}$$

The factor of 10^{-7} is included because electrical units are usually (MKS) and the acoustic units are (CGS).

The pressure produced at point B is related to the source pressure (p_0) produced at the reference distance r_0 through some function $g(r)$ which describes the spreading law**:

$$g(r_B)p_B = g(r_0)p_0$$

The pressure (p_F) produced at the transducer face is proportional to the strength of the source at point B:

$$|p_F| = \frac{h(f)}{g(r_B)}|Q| \tag{3.6}$$

where the spreading law is included and the factor h may be a function of frequency as indicated. These additional points are illustrated in Fig. 19.

---

* Or volume velocity, it has dimensions of L^3T^{-1}.

** For example, for spherical spreading $g(r) = 1/r$.

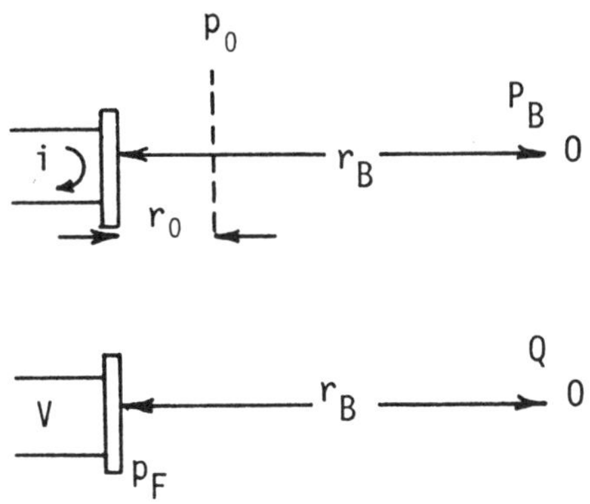

Fig. 19.

Putting these equations into (3.5) and rearranging results, we get:

$$\frac{|V|/|p_F|}{|p_0|/|L|} = \frac{g(r_0)}{h(f)} \cdot 10^{-7}$$

The right-hand side is defined to be the reciprocity parameter J, i.e.:

$$J = \frac{g(r_0)}{h(f)} \cdot 10^{-7} \quad (3.7)$$

and the left-hand side is the ratio of the receiver to transmitter responses, therefore:

$$\alpha/\beta = J \quad (3.8)$$

Note that J depends upon the transducer geometry through $h(f)$ and on the type of spreading through $g(r_0)$. J may be obtained directly from equation 3.6 with r_0 replacing r_B.

As an example of determining the reciprocity parameter, consider a planar transducer and a point close enough to the face so that the sound field may also be assumed to be planar. The reciprocity parameter is found using $p = \rho cu$ and $Q = 2Au$ (volume velocity with both faces moving). Thus we find for equation 3.6:

$$p = \frac{\rho c}{2A}(2Au) = \frac{\rho c}{2A} Q$$

Hence the planar reciprocity parameter is:

$$J_p = \frac{2A}{\rho c} \cdot 10^{-7}$$

The spherical reciprocity parameter is:

$$J_s = \frac{2\lambda r_0}{\rho c} \cdot 10^{-7} \tag{3.9}$$

Proving this will be left as a problem in the set following this chapter.

Transducer calibration by the reciprocity method requires three transducers: at least one reversible transducer (T), one projector (P), and one hydrophone (H). The three transducers are placed at the vertices of an equilateral triangle as shown in Fig. 20.

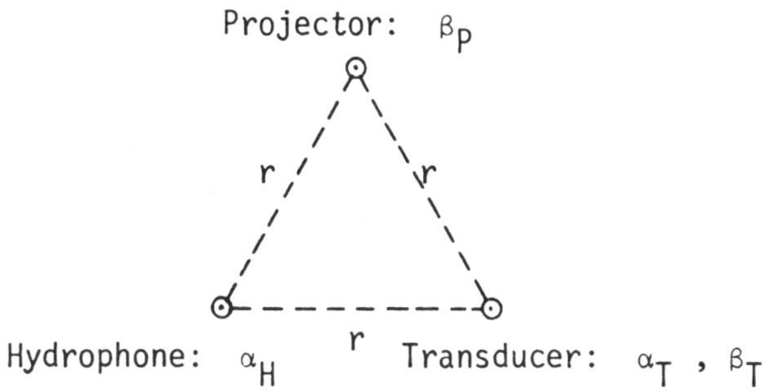

Fig. 20.

First, the projector is excited by a current i* and a voltage response V_H and V_T are measured at the hydrophone and the transducer, respectively. Then the transducer is excited by the same current i and the voltage response V'_H is measured at the hydrophone.

By reciprocity any or all of the responses, α_H, α_T, β_P, and β_T, can be calculated from the values of V_H, V'_H, V_T, r, i, and a knowledge of the type of spreading. The calculations are as follows:

By definition of the responses:

$$V_H = \alpha_H p_r; \quad V_T = \alpha_T p_r; \quad V'_H = \alpha_H p'_r \tag{3.10}$$

and:

$$p_{r0} = \beta_P i; \quad p'_{r0} = \beta_T i \tag{3.11}$$

By reciprocity:

$$\alpha_T = J\beta_T \tag{3.12}$$

Let us assume spherical spreading, $J = J_s$, then:

$$\beta_P = \frac{p_{r0}}{i} = \frac{p_r}{i}\left(\frac{r}{r_0}\right)$$

$$\beta_T = \frac{P'_{r0}}{i} = \frac{p'_r}{i}\left(\frac{r}{r_0}\right) \tag{3.13}$$

---

* Consider all currents, voltages and pressures mentioned here to be rms values.

By the first two equations of (3.10):

$$\frac{\alpha_H}{\alpha_T} = \frac{V_H}{V_T} \tag{3.14}$$

From equations 3.13:

$$\frac{\beta_P}{\beta_T} = \frac{p_r}{p_r'} \tag{3.15}$$

From the first and the last equations of (3.10):

$$\frac{\beta_P}{\beta_T} = \frac{p_r}{p_r'} = \frac{V_H}{V_H'} \tag{3.16}$$

From the products of equations 3.10 and 3.13:

$$\alpha_H \beta_P = \frac{p_r}{i}\left(\frac{r}{r_0}\right)\frac{V_H}{p_r} = \frac{V_H}{i}\left(\frac{r}{r_0}\right)$$

$$\alpha_H \beta_T = \frac{V_H'}{i}\left(\frac{r}{r_0}\right) \tag{3.17}$$

From the ratio of the first two equations of (3.10):

$$\alpha_H \beta_T = \alpha_T \frac{V_H}{V_T}$$

By reciprocity:

$$\alpha_H = J_s \beta_T \frac{V_H}{V_T} \tag{3.18}$$

By multiplying by α_H, it is found that:

$$\alpha_H^2 = J_s (\beta_T \alpha_H) \frac{V_H}{V_T}$$

Then, substituting from (3.17) results in:

$$\alpha_H^2 = J_s \frac{V_H'}{i}\left(\frac{r}{r_0}\right)\frac{V_H}{V_T}$$

Finally, the square root is taken:

$$\alpha_H = \left(J_s \frac{V_H'}{i}\frac{V_H}{V_T}\frac{r}{r_0}\right)^{1/2} \tag{3.19}$$

With this last equation and others above α_T, β_T, and β_P are readily obtained.

3.3. SPECIAL CALIBRATION TECHNIQUES

For accuracy in calibration it is necessary that the sound field be as free of reflections as is possible. This is achieved in either of two ways: large systems operated at low frequencies are calibrated at facilities where the reflecting surfaces can be kept as far removed as possible (i.e., deep lakes) and short pulses are used. In these cases the calibration data are taken on the direct pulse before the reflections arrive. The pulses must be greater in length than the reciprocal of the bandwidth of the system in order that a steady-state signal may be obtained, but short enough so that the reflections do not interfere. For smaller systems operated at higher frequencies, measurements are made in chambers with non-reflective walls (the reflected signals must be 20 to 40 dB below the direct). This non-reflective property of the wall is achieved by special design of the vessel geometry and wall liner material.

Broadband noise calibration techniques

A relatively quick and economical way to calibrate small hydrophones employs a tank flooded with broadband noise. The frequency response of a hydrophone is obtained by sweeping through the desired frequency range using a narrowband filter. Calibrations repeatable to within ±1 dB over frequencies down to about 200 Hz can be obtained for hydrophones smaller in linear dimension than about 10 inches.

Near-field calibration techniques

At a large distance from a projector (i.e., far field), the pressure varies as $1/r$. In the near field the pressure variation is more complicated. Roughly, the distance at which the transition to the far field begins is about a^2/λ where a is the maximum linear dimension of the transducer and λ is the wavelength of the sound projected. It is the far-field response of a transducer that is sought when the transducer is calibrated, but, because of space limitations, it is the near-field response that is measured when some very large transducers are calibrated.

The production of very large transducer arrays has prompted the development of near-field calibration techniques. Personnel of both the Underwater Sound Reference Laboratory, Orlando, Florida, and the Applied Research Laboratory*, Austin, Texas, have expended considerable effort on this problem.

---

* Formerly the Defense Research Laboratory.

The method developed at ARL* is based upon Green's theorem which states:

$$\int_S\int \left(\phi\frac{\partial\psi}{\partial n} - \psi\frac{\partial\phi}{\partial n}\right) dS = \int\int_V\int (\phi\nabla^2\psi - \psi\nabla^2\phi)dV$$

where $\phi(x, y, z)$ and $\psi(x, y, z)$ are the pressures at the point $P = (x, y, z)$ due to the transducer and a simple source, respectively. As a result of further development, it is found that the far-field pressure $p(P)$ is approximated by:

$$p(P) = -\frac{jk}{4\pi}\int_S\int (1 + \cos\beta)\frac{e^{jkr}}{r} p\, dS \qquad (3.20)$$

where β is the angle between the normal of the surface element and the line joining the surface element and the point P. In practice the calibration involves moving a small hydrophone probe over the surface S and measuring the amplitude and relative phase of the pressure at discrete points. The far-field pressure is then obtained from a numerical integration of (3.20). It has been found that accurate results can be obtained from a fairly small number of measurement points.

3.4. BEAM PATTERNS AND DIRECTIVITY

The receiving and transmitting responses of transducers are generally functions of spherical angles about the transducer. Let r represent either α or β and let it be a function of the spherical angles (θ, ϕ), i.e.:

$$r = r(\theta, \phi)$$

where ϕ is generally the angle for which there is the greatest symmetry. $r(0, 0)$ is the response of the transducer in the direction of the acoustic axis, and by definition of the acoustic axis $r(0, 0)$ is the maximum value of $r(\theta, \phi)$.

Beam pattern

Let us define the normalized response $\nu(\theta, \phi)$ by:

$$\nu(\theta, \phi) = \frac{r(\theta, \phi)}{r(0, 0)} \qquad (3.21)$$

so that $\nu(0, 0) = 1$. We define the beam pattern $b(\theta, \phi)$ by:

$$b(\theta, \phi) = \nu^2(\theta, \phi) \qquad (3.22)$$

---

* Defense Research Laboratory Report No. DRL-A-196, *The Determination of Farfield Characteristics of Large, Low-Frequency Transducers from Nearfield Measurements*, by D.D. Baker, 15 March 1962.

And finally, we define the log beam pattern by:

$$B(\theta, \phi)\ [\text{re}\ b(0,0) = 1] = 10 \log b(\theta, \phi)$$
$$= 20 \log \nu(\theta, \phi) \qquad (3.23)$$

Note that the beam pattern is always in the interval zero to one and the log beam pattern is always zero or less.

Although arrays will not be discussed until the next chapter, we will now derive a formula for the beam pattern of a discrete point line array. After the formula is obtained we will let the separation between elements go to zero and obtain the beam pattern for a continuous line transducer. For this development consider the diagram of Fig. 21.

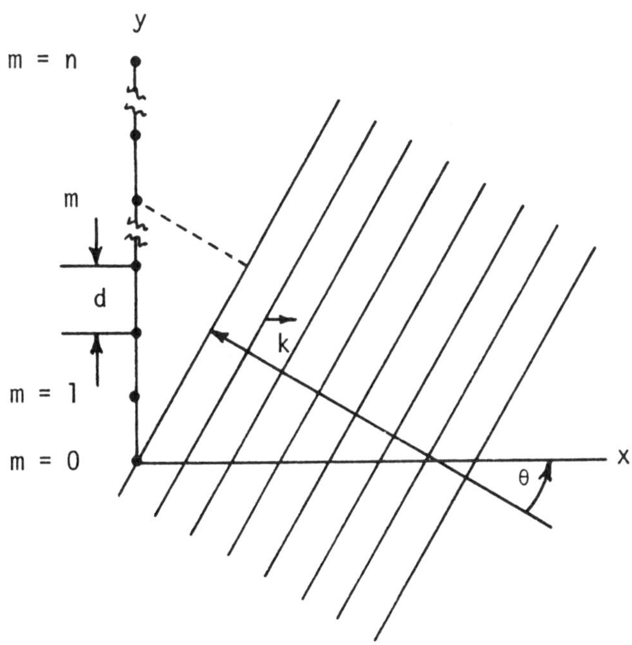

Fig. 21.

Assume that the plane wave:

$$p(x, y, t) = p_0 \exp[j(k \cdot r - \omega t)] = p_0 \exp[j(k_x x + k_z z - \omega t)]$$

is incident upon an array of $N(=n+1)$ hydrophones equally spaced along the z-axis ($x = 0$). The pressure along this axis is:

$$p_z = p_0 \exp[j(k_z z - \omega t)]$$

where $k_z = k \sin\theta = (2\pi/\lambda)\sin\theta$. The hydrophones are located at $z = md$ so that the pressure at the mth hydrophone is:

$$p_m = p_0 e^{j(mu-\omega t)}$$

where:

$$u = \frac{2\pi d}{\lambda}\sin\theta$$

The voltage at the mth hydrophone is:

$$V_m = \alpha_m p_0 e^{j(mu-\omega t)}$$

where α_m is the hydrophone's voltage response. We have neglected any relative phase shift between the pressure at the hydrophone and the voltage generated by it.

The output voltage is:

$$V = \sum_{m=0}^{n} V_m = p_0 e^{-j\omega t}\sum_{m=0}^{n}\alpha_m e^{jmu}$$

If the sensitivity is independent of m, then:

$$V = \alpha p_0 e^{-j\omega t}\sum_{m=0}^{n} e^{jmu}$$

It is readily shown that:

$$\sum_{m=0}^{n} e^{jmu} = \frac{e^{jNu}-1}{e^{ju}-1} = \exp\left[j\frac{(N-1)u}{2}\right]\frac{\sin Nu/2}{\sin u/2}$$

Therefore, the voltage is:

$$V = \alpha p_0 \exp\left\{j\left[\frac{(N-1)u}{2}-\omega t\right]\right\}\frac{\sin Nu/2}{\sin u/2} \tag{3.24}$$

The output is a sinusoidal voltage of the same frequency as the pressure wave, but shifted in phase by $(N-1)u/2$ relative to the phase of the first hydrophone.

The beam pattern is the normalized response squared. It may be calculated from the normalized mean square voltage, i.e.:

$$b(\theta) = \overline{V^2(\theta)}/\overline{V^2(0)}$$

where:

$$V(0) \equiv \alpha p_0 e^{-j\omega t}(N) \quad \text{and} \quad \overline{V^2} = VV^*$$

Then the beam pattern is:

$$b(\theta) = \left(\frac{\sin Nu/2}{N\sin u/2}\right)^2 \tag{3.25}$$

where $u = (2\pi d/\lambda)\sin\theta$. Note that $b(0) = 1$ and that the first zero in the beam pattern occurs at $Nu/2 = \pi$, i.e., for $\sin\theta = \lambda/dN$.

For a continuous line array, N goes to infinity as d goes to zero. These limits are taken in such a way that Nd goes to L, the length of the array. In this case:

$$b(\theta) = \left[\frac{\sin\left(\dfrac{\pi L}{\lambda}\sin\theta\right)}{\dfrac{\pi L}{\lambda}\sin\theta}\right]^2$$

A similar development for a circular plate transducer gives:

$$b(\theta) = \left[\frac{2J_1\left(\dfrac{\pi D}{\lambda}\sin\theta\right)}{\dfrac{\pi D}{\lambda}\sin\theta}\right]^2 \tag{3.26}$$

where D is the diameter of the plate and $J_1\,[(\pi D/\lambda)\sin\theta]$ is the first-order Bessel function. In this case $b(\theta)$ is also unity for $\theta = 0$, but the first zero occurs at:

$$\frac{\pi D}{\lambda}\sin\theta = 3.85$$

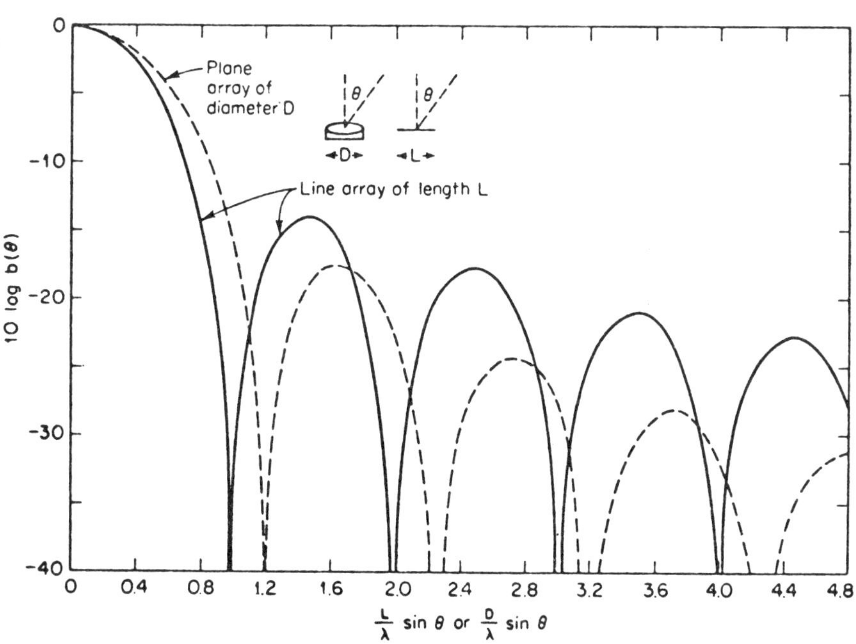

Fig. 22.

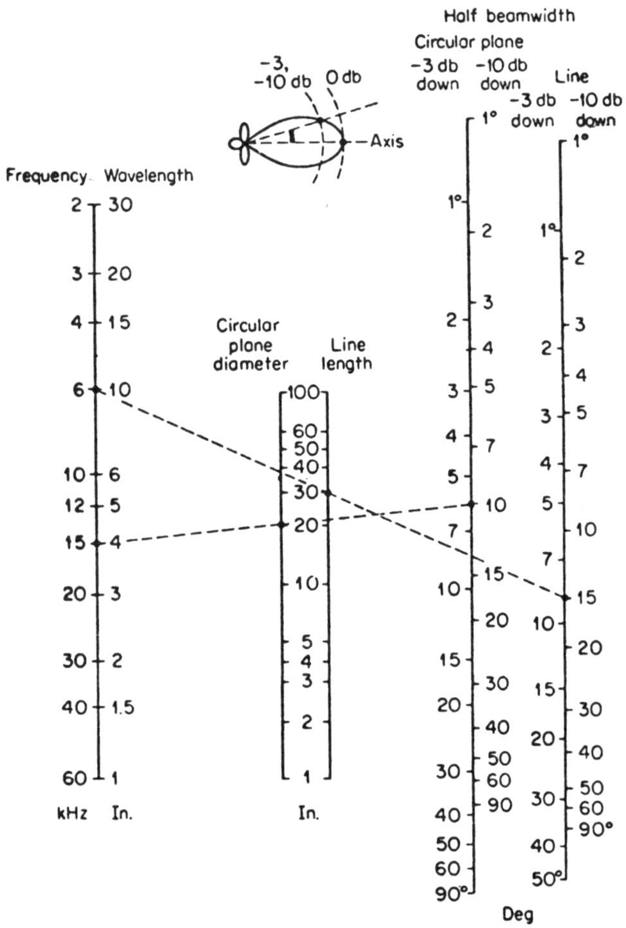

Fig. 23.

Figure 22 illustrates the beam patterns of a line and a circular plate transducer of length L and diameter D, respectively.* Figure 23 is a nomogram for finding the width of the beam pattern of line and circular plate transducers.*

* From *Principles of Underwater Sound for Engineers*, by R.J. Urick. McGraw-Hill, Inc., 1967. Used with permission of McGraw-Hill Book Company.

Receiver directivity index

Associated with the beam pattern is a quantity called the directivity index. This quantity is a measure of the ability of a transducer to discriminate against an isotropic noise in favor of a plane wave signal.

If $\overline{n^2}$ represents an isotropic noise power per unit solid angle, the mean square voltage produced by a non-directional hydrophone in the field is:

$$\overline{N^2_{\text{nond}}} = \text{constant} \int n^2 \mathrm{d}\Omega = 4\pi\overline{n^2}\,(\text{constant})$$

The mean-square voltage produced by the same hydrophone, with the exception that it is directional with beam pattern $b(\theta, \phi)$, is:

$$\overline{N^2_{\text{d}}} = \text{constant} \int \overline{n^2} b(\theta, \phi)\mathrm{d}\Omega = \text{constant}\,\overline{n^2} \int b(\theta, \phi)\mathrm{d}\Omega$$

The directivity index is defined by:

$$DI_{\text{R}} = 10 \log \frac{\overline{N^2_{\text{nond}}}}{\overline{N^2_{\text{d}}}} \tag{3.27}$$

so:

$$DI_{\text{R}} = 10 \log \frac{4\pi}{\int b(\theta, \phi)\mathrm{d}\Omega} \tag{3.28}$$

The meanings of θ and ϕ depend upon the general geometrical class of the transducer. For plane transducers the angles are standard spherical coordinate angles with the transducer lying in the z-plane. For linear transducers, the z-axis is taken along the transducer; ϕ is a standard spherical angle taken to be the angular measure around the transducer; and θ is measured from the z-plane (rather than from the z-axis).

Figure 24 illustrates these points.

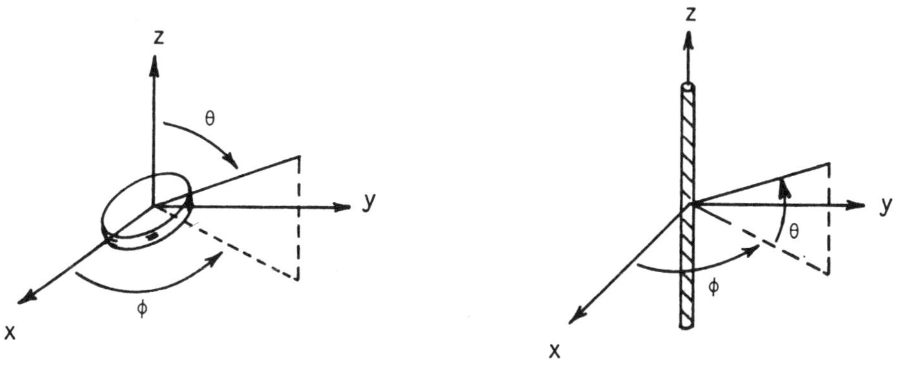

Fig. 24.

The integrals over the solid angles are different in these two cases: for the plane transducer:

$$\int b(\theta, \phi)\mathrm{d}\Omega = \int_0^{2\pi}\int_0^{\pi} b(\theta, \phi) \sin\theta \mathrm{d}\theta\mathrm{d}\phi$$

for the linear transducer:

$$\int b(\theta, \phi)\mathrm{d}\Omega = \int_0^{2\pi}\int_{\pi/2}^{\pi/2} b(\theta, \phi) \cos\theta \mathrm{d}\theta\mathrm{d}\phi$$

Transmitter directivity index

The transmitter directivity index is defined by:

$$DI_{\mathrm{T}} = 10 \log \frac{I_{\mathrm{S}}}{I_{\mathrm{nond}}} \tag{3.29}$$

where I_{S} is the source intensity of the projector with an output power P, and I_{nond} is the source intensity of a non-directional transmitter outputting the same power P. Figure 25 illustrates this.

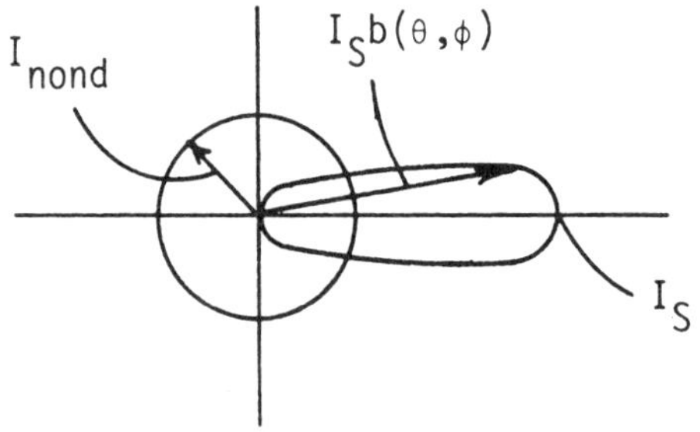

Fig. 25.

The power in the two separate cases is:

$$P = \int I_{\mathrm{S}}\, b(\theta, \phi)\mathrm{d}\Omega = I_{\mathrm{S}} \int b(\theta, \phi)\mathrm{d}\Omega^{*}$$

and:

$$P = \int I_{\mathrm{nond}}\, \mathrm{d}\Omega = 4\pi I_{\mathrm{nond}}$$

These are equal, so:

---

* Since the integral is over a sphere of unit radius, the radius need not appear in the equation.

$$4\pi I_{\mathrm{nond}} = I_{\mathrm{S}} \int b(\theta, \phi)\mathrm{d}\Omega$$

and:

$$\frac{I_{\mathrm{S}}}{I_{\mathrm{nond}}} = \frac{4\pi}{\int b(\theta, \phi)\mathrm{d}\Omega} \qquad (3.30)$$

The transmitter directivity index is:

$$DI_{\mathrm{T}} = 10 \log \frac{4\pi}{\int b(\theta, \phi)\mathrm{d}\Omega} \qquad (3.31)$$

This is the same as equation 3.28 for the receiver directivity index. That is, for the geometry and wavelength (i.e., the same beam pattern), the receiver and transmitter directivity indices are the same even though their initial definitions were different. Figure 26* is a nomogram giving the directivity indices for line and circular plate transducers.

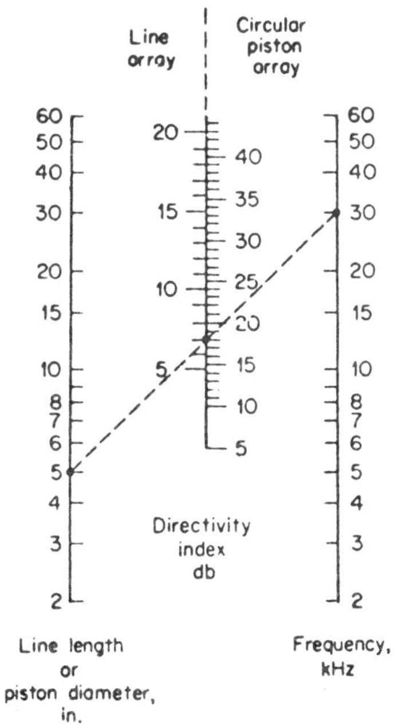

Fig. 26.

* From *Principles of Underwater Sound for Engineers,* by R.J. Urick. McGraw-Hill, Inc., 1967. Used with permission of McGraw-Hill Book Company.

Later we will have occasion to use the directivity factor (d) which we define by equation 3.30:

$$\frac{I_S}{I_{nond}} = \frac{4\pi}{\int b(\theta, \phi)\,d\Omega} = d \tag{3.32}$$

or by equation 3.31:

$$DI = 10 \log d \tag{3.33}$$

3.5. HYDROPHONE CHARACTERISTICS

Devices used in receiving underwater sound incorporate several elements: the sensor (hydrophone), the preamplifier (or transformer) and preamplifier housing, auxiliary circuits, and cables and cable connectors.

The sensor converts the sound pressure into an electrical voltage. Since the sensor is an electrostrictive element that is electrically equivalent to a capacitor with greater than 50,000 megohms of shunt resistance, a preamplifier or transformer must be used to lower the impedance before driving a long cable which generally has a great deal of capacitance (typically, 40 pF/ft). Auxiliary circuits are used for calibration, tests, and other purposes.

Hydrophones used in the ocean must be "seaworthy", which means that they must be capable of withstanding shipboard abuse, resist corrosion, biofouling, and extreme hydrostatic pressure. They must be reliable since regular inspections are most often impractical.

Generally, it is desired that the hydrophone be shielded from stray electrical fields. To accomplish this, often a screen-like metal grid surrounds the sensing element and the preamplifier is shielded. These shields are connected to a shielding on the cable which is grounded to instrument ground aboard ship. A sea ground is to be avoided because it may cause excessive ground-loop noise. This noise is caused by the large capacitance that exists between the shielding and the electrical components in the hydrophone. Modern hydrophones utilize completely independent internal shielding for the sensor element and the auxiliary circuits. This external shield is electrically insulated from the water-exposed metal housing.

Generally, preamplifier drivers for the cable are desired over inexpensive transformer-coupled systems because the latter have more limited frequency responses. Presently the most desired configuration is a low-noise field-effect-transistor (FET) preamplifier in proximity to the sensing elements. The preamplifier may or may not provide gain for its primary purpose is to reduce the impedance driving the lower end of the cable.

Well designed hydrophone systems provide a precision 10-ohm resistor in series with the sensor for remote calibration purposes. When this auxiliary circuit is used a separately shielded cable is required to drive this circuit in order to reduce crosstalk-feedback calibration error. Fig. 27 illustrates those points.

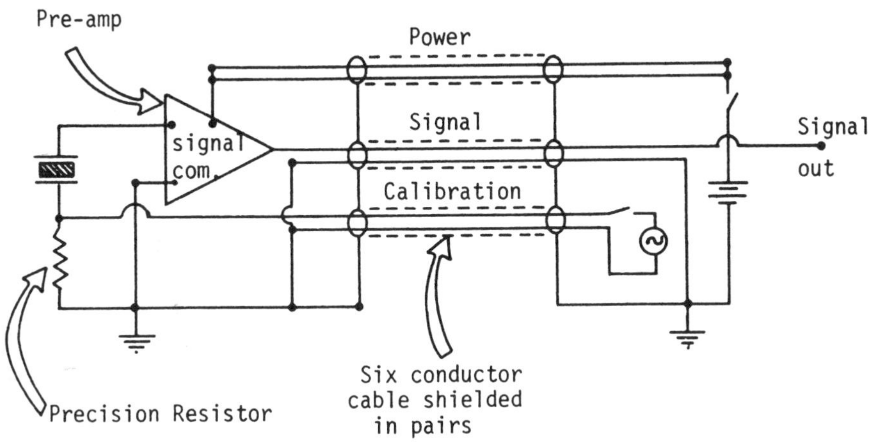

Fig. 27.

In addition to calibration, auxiliary circuits may be used to remotely control attenuation (to extend the dynamic range) and to limit the input signal in order to prevent overload caused by very strong acoustic signals.

3.6. PROJECTOR CHARACTERISTICS

The source intensity of a projector is determined by the acoustic power output and the directivity factor (or index):

$$P = 4\pi r_0^2 I_{\text{nond}} = 4\pi r_0^2 I_S / d$$

or:

$$I_{\text{nond}} = (4\pi r_0^2)^{-1} P = 0.95 \cdot 10^{-5} P$$

where $r_0 (= 1 \text{ yd} = 91.5 \text{ cm})$ is the reference distance for the source intensity. Therefore:

$$I_S = 0.95 \cdot 10^{-5} Pd$$

The "source level" is defined by:

$$SL = 10 \log (I_S / I_{\text{ref}}) \tag{3.34}$$

therefore:

$$SL = 10 \log (0.95 \cdot 10^{-5}) + 10 \log P + 10 \log d - 10 \log I_{\text{ref}}$$

Using equations 3.33 and 1.24 we find:

$$SL = 10 \log P + 71.5 + DI_T \tag{3.35}$$

Cavitation

As has been pointed out, piezoelectric crystals are generally limited in power output by dielectric breakdown and ferroelectric ceramics by mechanical breakdown. But it is possible in some cases for projectors made of either of these electrostrictive materials to experience an external limitation. This limitation is cavitation which is a mechanical breakdown of the medium itself.

Cavitation occurs when the net pressure (the instantaneous acoustic pressure plus the hydrostatic pressure) is less than the cohesive pressure of the medium. This cohesive pressure may be the vapor pressure of the medium itself or the pressure at which dissolved gases in the liquid form bubbles.

When cavitation occurs there is a loss of linearity in the transduction as the negative pressure half-cycles become "clipped", a reduction in directivity due to scattering from the bubbles, a loss in transmitted power due to coupling mismatch, and an increase in reverberation due to noise produced by the collapsing bubbles.

As an example of the latter, it may be shown that a gas (other than water vapor) filled bubble 1/10 cm^3 in volume collapsing to a pressure of 10 atm will produce 1/40 watt of power. A water vapor bubble is not so noisy because the vapor goes back into the liquid as the bubble collapses.

Let us define the cavitation threshold intensity (I_c in watt/cm^2) in terms of the peak negative pressure (p_c in atm) which the medium will support. Then:

$$I_c = \frac{[0.707 \cdot 10^6 p_c]^2}{\rho c} \cdot 10^{-7} = 0.3 p_c^2$$

where 10^6 is the conversion factor between atmospheres and dyn/cm^2 and the 0.707 is the conversion between peak and rms pressures. For example, the cavitation pressure of 1 atm is equivalent to the cavitation intensity of 0.3 W/cm^2 or 2 W/in^2.

The cavitation pressure varies with depth approximately as:

$$p_c(z) = p_c(0) + \frac{z}{33}$$

where z is the depth (positive downward) in feet, $p_c(0)$ is the cavitation pressure at the surface which is given by:

$$p_c(0) = 1 + T$$

where T is the tensile strength of the medium in atmospheres.

The cavitation intensity is given by:

$$I_c = 0.3\left[p_c(0) + \frac{z}{33}\right]^{2}{}^{*} \qquad (3.36)$$

The actual cavitation intensity may be as much as 50% lower than this value because in reality we may not have plane waves as is assumed here.

Cavitation appears to be dependent upon the history of the medium so that high-frequency and short-pulse projectors may operate at higher power than low-frequency and long-pulse projectors.

PROBLEMS

1. (a) What is the open-circuit response of a hydrophone that produces a rms voltage of 3 millivolts in a sound field having an rms pressure of 6 dyn/cm^2 (in decibels)?
   (b) What rms voltage is produced by a hydrophone with a response of -80 dB ref 1 volt per μbar in an rms pressure field of 0.6 dyn/cm^2?
   (c) What is the source level of a transducer having a transmitting current response of 96 dB and driven by 0.1 A peak-to-peak (twice amplitude) sinusoidal current?
2. Show that the spherical reciprocity parameter is given by equation 3.9.
3. Determine the relations for α_T, β_P, and β_T used in equations 3.10 and 3.11.
4. Show that:

$$V = \alpha p_0 e^{-j\omega t} \sum_{m=0}^{n} e^{jmu}$$

   reduces to equation 3.24. (Begin by multiplying the summation by $(e^{ju} - 1)$.)
5. What is the source level of a sonar operating with an input electrical power of 200 kW, directivity index of 20 dB, and 40% efficiency?

---

* Rule of thumb values for cavitation intensities are 1 W/cm^2 at the surface and 7 W/cm^2 at 100 feet.

CHAPTER 4

PHASED ARRAYS

Up to this point, with the exception of the development of equation 3.24, we have considered only single-element hydrophones and projectors and very simple arrays of non-shaded, non-phased, and non-directional elements. Many sonars in present operation use more complicated systems involving a variety of transducer elements electrically coupled in a variety of ways. In this chapter we shall discuss several of the more important aspects of these sonar arrays.

In the previous chapter the open-circuit voltage response of a hydrophone (or the pressure response of a projector) were considered to be real (i.e., phase shifts were neglected), and, for the array discussed, the response of each element was assumed to be non-directional and all responses were equal. In this chapter we shall see the effects of removing each of these restrictions.

First, let the response—we shall use the open-circuit voltage response (α) — be complex and defined by:

$$\underline{\alpha} = \alpha e^{j\delta}$$

where α is the magnitude as defined previously and δ is the phase shift between the cause (pressure) and the effect (voltage). Also let the response of each element in the array be, in general, different and directional, i.e.:

$$\underline{\alpha}_m = \alpha_m(\theta, \phi) \exp(j\delta_m)$$

where m labels the element.

A general expression for the voltage response of an additive array of N hydrophone elements to a sinusoidal plane wave is:

$$V = p_0 e^{-j\omega t} \sum_{m=0}^{N-1} \alpha_m \exp[j(\boldsymbol{k} \cdot \boldsymbol{r}_m + \delta_m)] \quad (4.1)$$

where $\boldsymbol{k}$ is the propagation vector and $\boldsymbol{r}_m$ is the position vector of the mth element.

The quantity $\boldsymbol{k} \cdot \boldsymbol{r}_m$ may be expanded by carrying out the dot product using $\boldsymbol{r}_m = (x_m, y_m, z_m)$ and $\boldsymbol{k} = (k\cos\theta, k\sin\theta\cos\phi, k\sin\theta\sin\phi)$ where k is the magnitude of $\boldsymbol{k}$ and θ and ϕ are standard spherical coordinate angles. But we are interested only in the fact that the dot product is a function of (θ, ϕ). Let:

$$\Delta_m(\theta, \phi) = k \cdot r_m$$

The beam pattern is defined by:

$$b(\theta, \phi) = \frac{\bar{V}^2(\theta, \phi)}{\bar{V}^2(0, 0)}$$

which reduces to:

$$b(\theta, \phi) = \left[\frac{\Sigma\alpha_m(\theta, \phi) \exp\{j(\Delta_m(\theta, \phi) + \delta_m)\}}{\Sigma\alpha_m(0, 0) \exp\{j(\Delta_m(0, 0) + \delta_m)\}}\right]^2 \tag{4.2}$$

We shall now discuss certain aspects of this formula.

Product theorem

The product theorem states: the overall beam pattern for an array of identical directional elements is the product of the beam pattern of an element and the beam pattern of the array of non-directional elements.

To demonstrate this theorem we use equation 4.2 but set:

$$\alpha = \alpha_1 = \alpha_2 = \ldots = \alpha(\theta, \phi)$$

and:

$$\delta = \delta_1 = \delta_2 = \ldots = 0$$

Then:

$$b(\theta, \phi) = \frac{\alpha^2(\theta, \phi)}{\alpha^2(0, 0)} \left[\frac{\Sigma \exp\{j\Delta_m(\theta, \phi)\}}{\Sigma \exp\{j\Delta_m(0, 0)\}}\right]^2$$

It is obvious that we can separate this into the part due to the directionality of the elements themselves ($b_e(\theta, \phi)$) and the part due to the geometry of the array ($b_a(\theta, \phi)$). Thus, the product theorem:

$$b(\theta, \phi) = b_e(\theta, \phi) \cdot b_a(\theta, \phi) \tag{4.3}$$

Electronic steering

An array may be steered, i.e., the acoustic axis turned to various directions, by either physically rotating the array or by providing prescribed phase shifts to the various elements; the latter is called "electronic steering". The general case of electronic steering is too complicated to be instructive, so we shall discuss the technique in terms of a simple case.

Consider the line array shown in Fig. 28.

If each element is phase-shifted such that:

$$\delta_m = \omega t_m = \omega S_m/c = kS_m = kmd \sin\theta_0$$

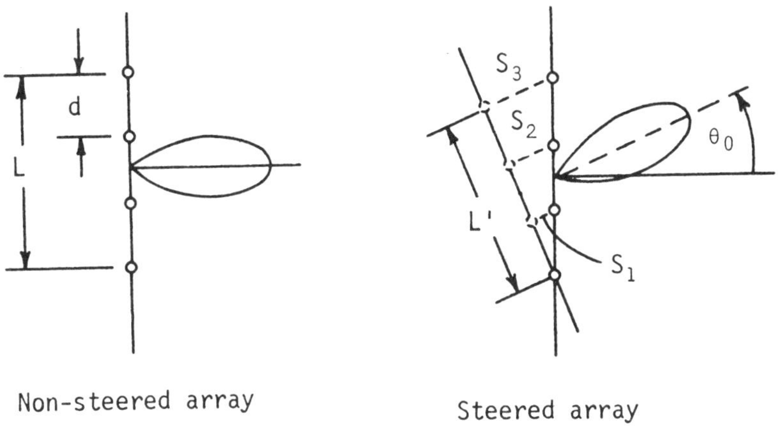

Fig. 28.

where ω is the angular frequency of the signal, t_m is the time delay introduced at the mth element, S_m is the effect spatial displacement along the line $\theta = \theta_0$, c is the speed of sound, and k is the wave number of the signal, then the line array will, in effect, be rotated by the angle θ_0, and the new beam pattern will be almost the same as the old but pointed in the new direction.

The beam patterns will not be exactly the same because of a modification of the length. The effective length of the electronically steared line array will be:

$$L' = L \cos \theta_0$$

Since the effective array length will be shorter than the actual length, the new beam pattern will be broader.

A more general discussion would begin with equation 4.2. Invariably, some approximations would be made so that a simple rotation of the original beam pattern will describe the new beam pattern for small rotations only. Steering through large angles will not only broaden the beam but, in general, distort it also.

Shading

Shading as applied to sonar arrays is a method by which some additional control of the beam pattern is obtained. Shading involves using a prescribed variation among the individual element responses. Making the array more responsive at the center than at the edges will broaden the beam but reduce the side lobes as compared to an unshaded array, and making the array more responsive at the ends will produce a narrower main beam but at the expense of increased side lobes. Two common forms of shaded linear arrays are the binomial shaded and the Dolph-Chebyshev shaded arrays.

Binomial shading involves taking the response of each element relative to the maximum response at the center to be in accordance with the coefficients of a binomial expansion of degree $(N-1)$ where N is the number of elements. For example, in the case of a six-element array the relative responses are 0.1, 0.5, 1.0, 1.0, 0.5, and 0.1. Binomial shading produces the narrowest main lobe with the total absence of side lobes.

The narrowest possible main lobe for a given side-lobe level is achieved through Dolph-Chebyshev shading. This involves shading in accordance with the coefficients of Chebyshev polynomials. For example, in the case of a six-element array the relative responses are 0.30, 0.69, 1.0, 1.0, 0.69, and 0.30.

Figure 29 illustrates the beam patterns of the unshaded, binomial shaded, and Dolph-Chebyshev shaded arrays of six elements:

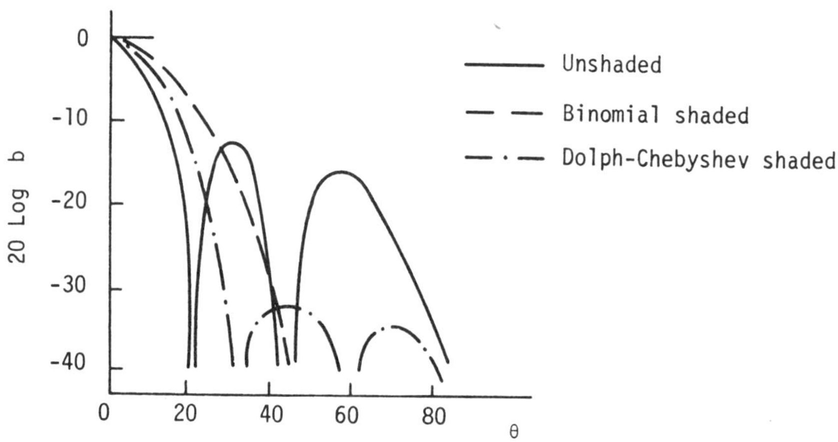

Fig. 29.

Array gain

The concept of receiver directivity index is applied to the situation of isotropic noise and plane wave signals. An extension to cover the more general situation brings us to the quantity array gain.

Let $s^2(\theta, \phi)$ and $n^2(\theta, \phi)$ represent the signal and noise powers per unit solid angle, respectively, as a function of the spherical angles, as indicated in Fig. 30. Let us adjust the signal level so that it is equal to the noise level when measured by a non-directional hydrophone, i.e.:

$$\overline{S^2} = \int_{4\pi} \overline{s^2}\, d\Omega = \int_{4\pi} \overline{n^2}\, d\Omega = \overline{N^2} \tag{4.4}$$

Therefore, the signal-to-noise ratio when measured by a non-directional hydrophone is 0 dB. We now define the array gain of a directional array

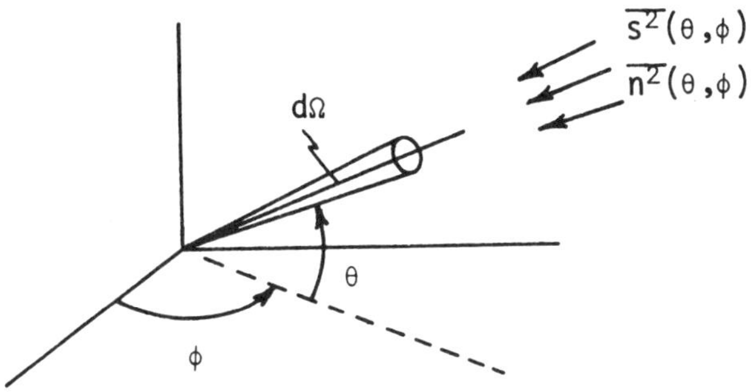

Fig. 30.

having a beam pattern $b(\theta, \phi)$ to be:

$$AG = 10 \log \frac{\int \overline{s^2}(\theta, \phi)\, b(\theta, \phi)\, d\Omega}{\int \overline{n^2}(\theta, \phi)\, b(\theta, \phi)\, d\Omega} \tag{4.5}$$

Therefore, the array gain is a measure of the signal-to-noise ratio of an array in decibels above a signal-to-noise ratio of zero as measured by a non-directional hydrophone. Naturally the array is designed to be better than a non-directional hydrophone so that AG is always a positive number of decibels. For a unidirectional signal in isotropic noise, array gain reduces to directivity index.

PROBLEMS

1. Show that in an isotropic noise field array gain of an array steered toward a unidirectional signal reduces to directivity index.
2. Since a product theorem applies to the beam pattern of an array, does a simple theorem (e.g., "addition theorem") apply to the directivity index? Discuss answer.

CHAPTER 5

SONAR EQUATIONS AND PARAMETERS

Let us begin our discussion of sonar equations with a rather simple acoustic situation and formulate the problem in terms of a multiplicative echo-ranging equation similar to that found in radar technology.

Given a source of source intensity I_S in a noiseless medium and a spherical target of radius a and reflectivity β, we seek the intensity of the echo I_e. Let the target be at some large range (relative to one yard) and assume that, due to spreading and attenuation, the intensity at the target will be reduced from the source intensity by the factor α. Then:

αI_S is the intensity at the target;

$\alpha I_S \pi a^2$ is the power intercepted by the spherical target;

$\beta \alpha I_S \pi a^2 / 4\pi R^2$ is the intensity of the reflected signal at the distance R from the center of the target along the axis joining the target center to the source (cf., equation 1.14; although the reflecting sphere, treated as a source, is not isotropic, it is approximately so along this axis);

$\beta \alpha I_S a^2/4$ is the reflected intensity at one yard from the target center; and

$\beta \alpha^2 I_S a^2/4$ is the echo intensity (I_e) received back at the source.

The result is, therefore:

$$I_e = \beta \alpha^2 I_S a/4$$

This is a multiplicative form of the echo-ranging equation for acoustics analogous to that used in radar.

Sonar technology uses an additive form of this equation obtained by converting to decibels, that is, divide by I_{ref}, take the logarithm, and multiply by 10:

$$10 \log \frac{I_e}{I_{ref}} = 10 \log \left[\frac{\beta \alpha^2 a^2}{4} \left(\frac{I_S}{I_{ref}} \right) \right]$$

This results in the echo level (EL):

$$EL = SL + 20 \log \alpha + 10 \log \beta a^2/4$$

The quantity $-10 \log \alpha$ is the one way transmission loss (TL) and $10 \log \beta a^2/4$ is the target strength (TS) for a sphere of radius a and reflectivity β. A more general definition of target strength will be given later but it should be pointed out with this simple example that it is a positive quantity when $\beta \alpha^2 > 4$ (in the case of unit reflectivity, when $a > 2$ yd).

The equation for the echo level in this simple case is:

$$EL = SL - 2TL + TS \tag{5.1}$$

But the equation is quite general and applies to any echo-ranging problem for which the transmitter and receiver are quite close together.

The transmission loss term includes signal reduction due to spreading and attenuation (absorption and scattering). In the case of spherical spreading (inverse square law) with no attenuation:

$$TL = -10 \log (1/r^2) = 20 \log r$$

where r is in yards to be consistent with the previous definition of source level.

5.1. VARIOUS FORMS OF SONAR EQUATIONS

The basic problem in sonar is to measure some signal (possibly an echo) against a background of noise (or reverberation). In order that the signal may be detected above the background, the ratio of the measured signal to the measured background ("signal-to-noise ratio") must be at least some minimum value that is determined by the system.

Let DT (detection threshold) be the minimum detectable signal level (MDS) when the noise level (NL)* is zero decibels, that is, DT is the minimum detectable signal-to-noise ratio. We express this by the following equation:

$$DT = MDS - NL \tag{5.2}$$

Generally, the design criterion for a functional sonar can be expressed by the following inequality:

$$\text{SIGNAL} \geqslant MDS = DT + NL \tag{5.3}$$

The basic minimal operational condition for sonars is represented by equation 5.2 and may be illustrated with certain cases:

— The signal is the echo level given by equation 5.1, the noise is isotropic and of level NL, and the receiving hydrophone is non-directional; equation 5.2 becomes:

$$SL - 2TL + TS - NL = DT \tag{5.4}$$

In this case the system is said to be noise-limited.

— The system is same as above except that the hydrophone is directional with directivity index DI; equation 5.2 becomes:

---

* In this equation and those that follow NL is the spectrum level. The $10 \log W$ that should be present is absorbed into DT.

$$SL - 2TL + TS - NL + DI = DT \tag{5.5}$$

— The system is the same as above except that the transmitter is putting so much energy into the water that the reverberation level RL (the level of the backscattered source energy as measured by the hydrophone) exceeds the measured noise level; equation 5.2 becomes:

$$SL - 2TL + TS - RL + AG = DT \tag{5.6}$$

where AG is array gain which must be used instead of directivity index since RL is not isotropic. In this case the system is said to be reverberation-limited. Some advantage is gained in a noise-limited system by increasing the output power, if feasible, until the system becomes reverberation-limited.

— The system is one in which the transmitter and receiver are separated by an appreciable distance (bistatic echo-ranging sonar). The transmission losses to and from the target are not, in general, equal; equation 5.2 becomes:

$$SL - TL_1 - TL_2 + TS - NL + DL_T + DI_R = DT \tag{5.7}$$

This assumes that the projector and receiver are pointed at the target.

— The system is passive and the hydrophone is pointed at the target of target source level SL; equation 5.2 becomes:

$$SL - TL - NL + DI = DT \tag{5.8}$$

These equations are used in the design or performance prediction of a sonar system. For a given sonar system with a specified detection threshold acting against a target of known strength in a known noise environment, one may seek the source level needed to attain a certain range or the range attained for a given source level.

The previous discussion of the sonar equations is valid for long pulses for which the effective emitted and received pulse durations are about the same. If this is not the case a modification of the source level is required. The source level SL must be replaced by the effective source level SL' defined by:

$$SL' = SL + 10 \log \tau_0/\tau_e \tag{5.9}$$

where τ_0 and τ_e are emitted and echo pulse durations, respectively.

5.2 SONAR PARAMETERS AND THEIR VARIOUS COMBINATIONS

The various terms in the sonar equations are called sonar parameters. For their definitions and reference locations see Urick (1967), p. 21.

These parameters may be grouped according to whether they are determined by the equipment, medium, or target. This grouping is as follows:

Equipment Parameters
SL: Source Level
DT: Detection Threshold
DI: Directivity Index
NL: Self-Noise Level
AG: Array Gain (also determined by medium)
Medium Parameters
TL: Transmission Loss
NL: Ambient Noise Level
RL: Reverberation Level (also determined by equipment)
Target Parameters
TS: Target Strength
SL: Target Source Level

Oceanographic interest in marine acoustics is concentrated mostly in the parameters determined by the medium.

Various combinations of the sonar parameters have also been given names, for example, as we have already seen $SL - 2TL + TS$ is the echo level, $NL - DI$ is the noise measured at the hydrophone terminals, and SIGNAL $- MDS$ is the echo excess. A commonly used combination of sonar parameters is the figure of merit (*FOM*). It is defined by:

$$FOM = SL - NL + DI - DT \tag{5.10}$$

The figure of merit equates to the allowable transmission loss.

PROBLEMS

1. Given an active monostatic sonar having a source level of 115 dB, a receiving directivity index of 12 dB, a processor giving a detection threshold of 20 dB, and assuming a noise (spectrum) level of −40 dB and spherical spreading, find the maximum range at which a target of strength 15 dB may be detected.
2. We would like to study the deep scattering layer with a system employing a 100 watt power amplifier, a projector that is 40% efficient, a receive and transmit directivity index of 5 dB each, and a processor giving a detection threshold of 23 dB. Assume that a *DSL* of target strength equal to −40 dB is at a depth of 400 yd, the spreading is spherical, and the noise (spectrum) level is −50 dB. Can the *DSL* be detected in this environment with this system and what is the echo excess or deficit?
3. We would like to design a shallow-water transmission experiment. All the parameters of the system are fixed except the power. These parameters are: projector and receiver directivity index, 22 dB; projector efficiency, 30%; detection threshold, 10 dB. Assume a background noise (spectrum) level of −30 dB, cylindrical spreading and attenuation of 1 dB/kyd. What

is the minimum power required from the amplifiers in order that a signal may be detected at a range of 10 nt. miles?

4. A long-pulse sonar of source level 130 dB, receiving directivity index of 10 dB echo ranges against a target of strength 20 dB. In a noise background spectrum level of −30 dB and with *DT* of +10 dB, and if spherical spreading is assumed, what is the maximum range for detection?
5. A fish-finding sonar is to be designed to detect a fish school of target strength 0 dB, at a range of 1000 yd. It is to operate at 50 kHz, have a source level of 100 dB and the echo must be 10 dB (*DT*) above the background noise of −40 dB. Assume spherical spreading with an additional absorption loss of 15 dB. How large a diameter must a circular plate transducer have in order to just accomplish this? Considering cavitation, would you think that this system would function with the transducer near the surface?

CHAPTER 6

THEORY OF SOUND PROPAGATION

According to a simple theory, sound propagates in the sea in a manner that is controlled by a linear, second-order partial differential equation known as the "wave equation". This equation is obtained from four basic equations:

(1) The equation of continuity (the mathematical expression for the law of conservation of mass):

$$\frac{\partial \rho}{\partial t} + \nabla \cdot (\rho \boldsymbol{u}) = 0 \tag{6.1}$$

where ρ is the density of the fluid, and $\boldsymbol{u}$ is the particle velocity.

(2) Equation of motion (Newton's second law as applied to small volumes of a fluid):

$$f = \frac{\partial}{\partial t}(\rho \boldsymbol{u}) \tag{6.2}$$

where f is the force per unit volume.

(3) Force—pressure relation:

$$f = -\nabla p \tag{6.3}$$

(4) The equation of state (which, in this approximate theory, is taken to be an expression of Hooke's Law):

$$p_e = \frac{k}{\rho_0} \rho_e \tag{6.4}$$

where $p = p_0 + p_e$ and $\rho = \rho_0 + \rho_e$. Subscripts 0 and e stand for equilibrium and excess, respectively, and k is the bulk modulus. It is assumed that the equilibrium values are independent of space and time.

The wave equation is obtained from the above equations as follows: The force—pressure relation and the equation of motion are combined to obtain:

$$-\nabla p = \frac{\partial}{\partial t}(\rho \boldsymbol{u})$$

(This equation may also be called the equation of motion). The divergence of this equation is taken, the order of the differential operators on the right is changed and the equation of continuity is used to obtain:

$$\nabla^2 p = \frac{\partial^2 \rho}{\partial t^2}$$

where ∇^2 is the Laplacian operator. Since the equilibrium values are assumed to be uniform in space and constant in time, this equation reduces to:

$$\nabla^2 p_e = \frac{\partial^2 \rho_e}{\partial t^2}$$

Using equation 6.4 and the definition:

$$c^2 = k/\rho_0 \tag{6.5}$$

where c will later be shown to be the sound speed, the above equation reduces to the wave equation in either of two forms:

$$\nabla^2 \rho - \frac{1}{c^2}\frac{\partial^2 \rho}{\partial t^2} = 0 \tag{6.6}$$

and:

$$\nabla^2 p - \frac{1}{c^2}\frac{\partial^2 p}{\partial t^2} = 0 \tag{6.7}$$

where p and ρ, without subscripts, now, and throughout the remainder of these notes, mean the excess of "acoustical" pressure and density, respectively.

These wave equations are solved by two basic approaches: wave theory and ray theory. In the wave-theory approach functional solutions of the linear, second-order partial differential equation and a set of boundary conditions are sought using standard techniques. In the ray-theory approach a specific form for the solution is assumed and inserted into the wave equation to obtain another equation known as the "eikonal equation".

6.1. WAVE THEORY

The general solution of the wave equation with constant c is:

$$p(r, t) = f(k \cdot r - \omega t) + g(k \cdot r + \omega t)$$

where $c = \omega/k$. Or in one dimension:

$$p(x, t) = f(x - ct) + g(x + ct)$$

This solution is easily verified by direct substitution into the wave equation.

If either g or f is zero, the solution represents a rigid wave form propagating in the direction of the positive or negative x-axis, respectively. This is readily shown; for example, by taking $g = 0$ and asking when will $f(x_1 - ct_1)$ be identically equal to $f(x_2 - ct_2)$. The answer is when:

$$x_1 - ct_1 = x_2 - ct_2$$

or when:

$$c = \frac{x_2 - x_1}{t_2 - t_1}$$

Thus showing that c is the speed of propagation and the motion is toward the positive x-axis.

The same may be done with $f = 0$. In that case we find that $g(x + ct)$ represents a wave form propagating with the speed $-c$, i.e., in the direction of the negative x-axis. If both f and g are non-zero, then no rigid profile exists and no specific direction or speed of propagation can be assigned to the motion of the total profile. If f and g are identical, then a standing wave exists.

To obtain the actual functions that satisfy the wave equation it is necessary to have boundary and initial conditions in addition to the wave equation itself. We will consider two methods of obtaining solutions to specific boundary value problems involving the wave equation.

The method of separation of variables with application to the wave equation in one dimension

Suppose we wish to solve the wave equation in the form:

$$\frac{\partial^2 p}{\partial x^2} - \frac{1}{c^2}\frac{\partial^2 p}{\partial t^2} = 0 \tag{6.8}$$

subject to the boundary conditions:

$$p(0, t) = 0 \quad \text{and} \quad \frac{\partial p}{\partial x}(L, t) = 0^* \tag{6.9}$$

and the initial conditions:

$$p(x, 0) = f(x) \quad \text{and} \quad \frac{\partial p}{\partial t}(x, 0) = g(x) \tag{6.10}$$

The above example describes a "boundary value problem" the solution of which we may obtain by the method of separation of variables, viz., we try a solution in the form:

$$p(x, t) = X(x)\,T(t)$$

* These boundary conditions are appropriate for acoustic propagation in a layer with acoustically soft (e.g., water—air) and hard (water—rock) interfaces, respectively.

This trial solution is substituted back into the partial differential equation and the result is divided by the solution to obtain:

$$\frac{X''(x)}{X(x)} = \frac{1}{c^2}\frac{T''(t)}{T(t)}$$

where the primes represent differentiation by the argument of the function. We see that the *RHS* is a function of t alone and the *LHS* of x alone; therefore, they must both be constant. Set them equal to a constant $-k^2$, so that:

$$X''(x) + k^2X(x) = 0 \tag{6.11}$$

and:

$$T''(t) + \omega^2T(t) = 0 \tag{6.12}$$

where $\omega^2 = k^2c^2$.

The boundary conditions become:

$$X(0) = 0, \quad \text{and} \quad X'(L) = 0 \tag{6.13}$$

Equations 6.11 and 6.12 have the solutions:

$$X(t) = A\mathrm{e}^{jkx} + B\mathrm{e}^{-jkx} \tag{6.14}$$

and:

$$T(t) = C\mathrm{e}^{j\omega t} + D\mathrm{e}^{-j\omega t} \tag{6.15}$$

For equation 6.14 to satisfy the first boundary condition of (6.13), it is necessary that:

$$A + B = 0$$

So we let:

$$\frac{1}{2j}a = A = -B$$

Then:

$$X(t) = \frac{1}{2j}a(\mathrm{e}^{jkx} - \mathrm{e}^{-jkx}) = a \sin kx \tag{6.16}$$

And to satisfy the second condition it is necessary that:

$$X'(L) = -ak \cos kL = 0$$

therefore, k has discrete values, k_n, which satisfy:

$$k_n = \frac{2n+1}{2}\frac{\pi}{L}, \quad n = 0, 1, \ldots, \infty \tag{6.17}$$

Let us digress a moment. We have the differential equation:

$$X_n'' + k_n^2X_n = 0 \tag{6.18}$$

where the X_n is a solution given by:

$$X_n = a_n \sin k_n x \qquad (6.19)$$

corresponding to the nth value of k which is given by equation 6.17. The above problem with its boundary conditions forms an eigenvalue problem with eigenvalues k_n and eigenfunctions X_n (solutions called normal modes).

To analyze the time dependence we recognize that we must label the solutions of equation 6.15 by n to correspond with the nth value of ω which is equal to $k_n c$.* An alternative way of writing the solution is:

$$T_n(t) = \cos(\omega_n t + \theta_n)$$

where the coefficient is suppressed since it may be included in a_n.

Normal modes satisfying the wave equation and boundary conditions are:

$$p_n(x, t) = a_n \sin k_n x \cos(\omega_n t + \theta_n)$$

The general solution may be formed as a sum of all the normal modes, i.e.:

$$p(x, t) = \sum_{n=0}^{\infty} a_n \sin k_n x \cos(\omega_n t + \theta_n) \qquad (6.20)$$

This solution satisfies the initial conditions provided:

$$f(x) = \sum_{n=0}^{\infty} a_n \cos\theta_n \sin k_n x$$

and:

$$g(x) = -\sum_{n=0}^{\infty} \omega_n a_n \sin\theta_n \sin k_n x$$

The set $[\sin\{(2n+1)\pi x/2L\}]$ is orthogonal in the interval $[0, L]$ with normalization:

$$N = L/2^{**} \qquad (6.21)$$

---

* $k_n c_n$ actually, but we assume that the speed of acoustic propagation is independent of the wave number.

** Proof:

$$\int_0^L \sin\left[\frac{(2k+1)\pi x}{2L}\right] \sin\left[\frac{(2j+1)\pi x}{2L}\right] dx = \int_0^L \sin\frac{m\pi x}{L} \sin\frac{n\pi x}{L} dx$$

where: $m = \dfrac{2k+1}{2}, \quad n = \dfrac{2j+1}{2}$

This is: $\dfrac{\sin(m-n)\pi}{\pi(m-n)/L} - \dfrac{\sin(m+n)\pi}{\pi(m+n)/L} = 0 \quad m^2 \neq n^2$

or: $N = \displaystyle\int_0^L \sin^2\frac{m\pi x}{L} dx = \frac{L}{2} \quad m = n$

Therefore, the coefficients (a's) and the phases (θ's) may be determined from:

$$a_n \cos \theta_n = \frac{1}{N} \int_0^L f(x) \sin \frac{(2n+1)\pi x}{2L} \, dx$$

and:

$$a_n \sin \theta_n = -\frac{1}{\omega_n N} \int_0^L g(x) \sin \frac{(2n+1)\pi x}{2L} \, dx \qquad (6.22)$$

These last equations may be solved to give a_n and θ_n separately.

Figure 31 illustrates several normal modes of vibration consistent with this problem.

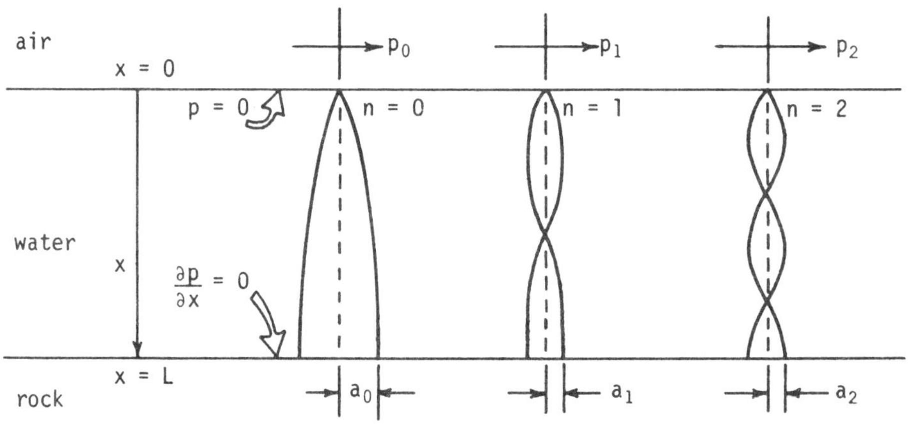

Fig. 31.

The Green's function method

The method of separation of variables works in only a few coordinate systems. What is really needed is a method of solving the wave equation which would apply to all boundary shapes. The formulation in terms of the Green's functions is a step in that direction. We shall develop the Green's function method and use it to develop the Helmholtz integral formula for a field set up by a point source in a certain bounded space.

Let us attempt to solve the wave equation in the form:

$$\nabla^2 \phi' - \frac{1}{c^2} \frac{\partial^2 \phi'}{\partial t^2} = -4\pi\psi' \qquad (6.23)$$

where $\phi' = \phi'(r, t)$ is the velocity potential, pressure, or density, and $\psi' = \psi'(r, t)$ is a source function for acoustic energy. Let us assume a harmonic time dependence* for ϕ' and ψ' of the form:

$$\phi'(r, t) = \phi(r)\, e^{-j\omega t}$$

$$\psi' = \psi(r) e^{-j\omega t}$$

In this case equation 6.23 reduces to:

$$\nabla^2\phi + k^2\phi = -4\pi\psi \tag{6.24}$$

where $k^2 = \omega^2/c^2$.

Let us also consider the function $G(r, r_0)$ called "Green's function" which satisfies:

$$\nabla^2 G + k^2 G = -4\pi\delta(r - r_0) \;\; ^{**} \tag{6.25}$$

and certain boundary conditions determined by the problem. Green's function is the field set up by a point source in a space with boundary surfaces consistent with the problem at hand.

We now multiply equation 6.24 by G, equation 6.25 by ϕ, and subtract to get:

$$G\nabla^2\phi - \phi\nabla^2 G = -4\pi G\psi + 4\pi\phi\delta(r - r_0) \tag{6.27}$$

We now integrate over a volume V enclosing r_0 to obtain:

$$\iiint_V (G\nabla^2\phi - \phi\nabla^2 G)\,\mathrm{d}V = -4\pi \iiint_V G(r, r_0)\,\psi(r)\,\mathrm{d}V + 4\pi\phi(r_0)$$

We may write the *LHS* as:

$$\iiint_V \nabla \cdot (G\nabla\phi - \phi\nabla G)\,\mathrm{d}V$$

as may be verified by carrying out the operation. Then, by the divergence theorem we may write the *LHS* as a surface integral:

$$\iint_S (G\nabla\phi - \phi\nabla G) \cdot \boldsymbol{n}\,\mathrm{d}S$$

---

* Not particularly restrictive since we may eventually sum over the frequency components.

** $\delta(r - r_0)$ is the Dirac delta function defined by:

$$(1)\;\delta(r - r_0) = \begin{cases} 0 & r \neq r_0 \\ \infty & r = r_0 \end{cases}$$

$$(2)\iiint_V \delta(r - r_0)\,\mathrm{d}V = \begin{cases} 0 & \text{if } r_0 \text{ not in } V \\ 1 & \text{if } r_0 \text{ in } V \end{cases}$$

$$(3)\iiint_V f(r)\delta(r - r_0)\,\mathrm{d}V = \begin{cases} 0 & \text{if } r_0 \text{ not in } V \\ f(r_0) & \text{if } r_0 \text{ in } V \end{cases} \tag{6.26}$$

where n is the outward normal on the surface. We then write the *LHS* as:

$$\int\int_S \left(G\frac{\partial \phi}{\partial n} - \phi\frac{\partial G}{\partial n}\right) \mathrm{d}S$$

where $\partial/\partial n$ is the normal derivative at the surface. Putting this back into the equation, solving for $\phi(r_0)$, and letting r_0 become r and r become r' we find:

$$\phi(r) = \int\int\int_V G(r, r')\psi(r')\mathrm{d}V' + \frac{1}{4\pi}\int\int_S \left(G\frac{\partial \phi}{\partial n'} - \phi\frac{\partial G}{\partial n'}\right) \mathrm{d}S' \qquad (6.28)$$

where r represents an observation point.

If there are no sources in V (i.e., $\psi = 0$ for all r in V), the volume integral is zero, or, if there are no surfaces in the problem, the surface intergrals are zero. The two integrals are not independent and one may be eliminated by choosing either of the boundary conditions:

$$G = 0 \text{ on } S \quad \text{if } \phi \text{ is known on } S$$

or:

$$\frac{\partial G}{\partial n} = 0 \text{ on } S \quad \text{if } \frac{\partial \phi}{\partial n} \text{ is known on } S$$

Green's function is obtained by solving (6.25) with the appropriate boundary condition.

We conclude therefore that, if the appropriate Green's function, source distribution, and boundary conditions are known, a complete solution of the boundary-value problem may be obtained.

This is a perfectly general approach and, in theory, will solve many wave problems that separation of variables can not. But, in practice, analytical solutions may be obtained for simple geometries only because Green's functions can not be obtained analytically for complicated surfaces. However, numerical solutions have been obtained for complicated geometries using the Green's function method.

The Helmholtz integral formula

It may be shown that the "free space Green's function":

$$G(r, r') = \frac{\exp(jk|r - r'|)}{|r - r'|} = \frac{e^{jkR}}{R} \qquad (6.29)$$

satisfies equation 6.25. While this Green's function is particularly suited for use in unbounded space, it may be used in any space. However, it satisfies no particular boundary condition so all terms in equation 6.28 must be used.

With this Green's function equation 6.28 reduces to:

$$\phi(r) = \frac{1}{4\pi}\int\int_S \left[\frac{e^{jkR}}{R}\frac{\partial\phi(r')}{\partial n'} - \phi(r')\frac{\partial}{\partial n'}\left(\frac{e^{jkR}}{R}\right)\right] dS' + \int\int\int_V \psi(r')\frac{e^{jkR}}{R}\, dV' \quad (6.30)$$

If there are no sources in V (i.e., $\psi(r') = 0$ for all r' in V) then:

$$\phi(r) = \frac{1}{4\pi}\int\int_S \left[\frac{e^{jkR}}{R}\frac{\partial\phi(r')}{\partial n'} - \phi(r')\frac{\partial}{\partial n'}\left(\frac{e^{jkR}}{R}\right)\right] dS' \quad (6.31)$$

The above equation is known as the "Helmholtz integral formula", and is used to determine the field in an enclosed volume containing no sources while the values of ϕ and its normal derivative are known on the enclosing surface.

The eikonal equation

Ray theory, good for high frequencies, is based upon an equation called the "eikonal equation". The eikonal equation is obtained from the wave equation by assuming a solution of a particular form and substituting the solution into the wave equation.

The solution is assumed to be:

$$\phi = A\exp[j(k_0 W - \omega t)] \quad (6.32)$$

where A and W are functions of the position vector r, and k_0 is a constant. When this is put into the wave equation and the indicated operations are performed, one finds:

$$\nabla^2 A - k_0^2 A\,|\nabla W|^2 + \frac{\omega^2}{c^2}A + 2jk_0\nabla A\cdot\nabla W + jk_0 A\nabla^2 W = 0$$

For the equality to be valid, it is necessary that both the real and the imaginary parts be zero:

$$|\nabla W|^2 - \frac{\omega^2}{k_0^2 c^2} = \frac{\nabla^2 A}{k_0^2 A} = \frac{\lambda_0^2\nabla^2 A}{4\pi^2 A} \quad (6.33)$$

and: $A\nabla^2 W + 2\nabla A\cdot\nabla W = 0$ (6.34)

If the *RHS* of (6.33) is small (that is, if A is not a strong function of the coordinates and if λ_0 is small), the equation becomes:

$$|\nabla W|^2 - (c_0^2/c^2) = 0 \quad \text{or} \quad |\nabla W|^2 - n^2 = 0 \quad (6.35)$$

where $c_0^2 = \omega^2/k_0^2$ and $n^2 = c_0^2/c^2$. n is the index of refraction and may be dependent upon the position vector r.

Equation 6.35 is called the eikonal equation and W is called the eikonal. The eikonal $W(r)$ is a surface in three-dimensional space that can be associated with wave fronts (i.e., surfaces of constant phase). This is readily seen if the phase of (6.32) is equated to a constant at a particular time (t_0):

$$\omega t_0 - k_0 W = \text{constant}$$

Then:

$$W(r) = \frac{\omega t_0 - \text{constant}}{k_0} \tag{6.36}$$

which describes a surface in space. Ray trajectories are everywhere perpendicular to the wave fronts and hence are also determined by equation 6.36.

It should be noted that the well known W.K.B. approximation is a special case of the eikonal method. The W.K.B. approximation applies to the case of a vertically stratified medium where A and W are functions of z only.

Fermat's principle

An equivalent way of formulating ray theory is based on Fermat's principle for the trajectory of rays. This principle states that the path a ray will take in going from point A to point B (see Fig. 32) is such that the time:

$$t = \int_A^B \frac{ds}{c(s)}$$

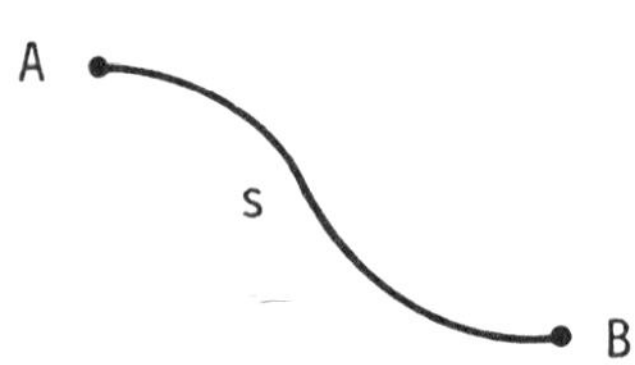

Fig. 32.

where $c(s)$ is the speed of sound as a function of the parameter s, will be an extremum.

Snell's law

Both the eikonal method and Fermat's principle lead to Snell's law in a continuum; that is:

$$\frac{c}{\cos\theta} = \frac{c'}{\cos\theta'} = c_v \tag{6.37}$$

where c_v is a constant for a given ray and θ is an angle as shown in Fig. 33.

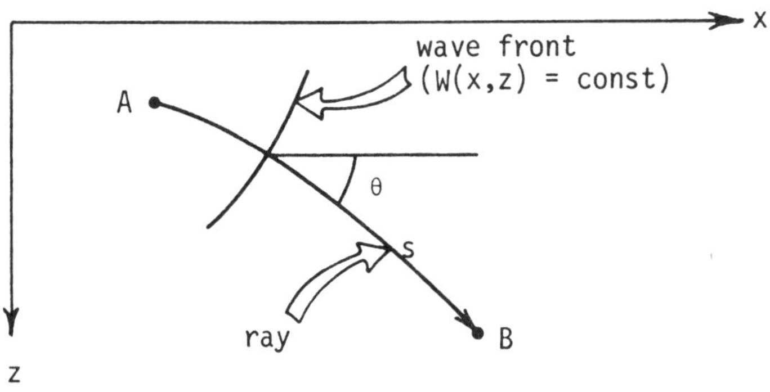

Fig. 33.

The physical significance of c_v may be seen by letting $\theta = 0$. Thus, we see that c_v is the velocity at the point where the ray becomes horizontal and therefore is different from one ray to the next. This point is called the "vertex" of the ray and c_v is called the "vertex velocity".

The ray method determines the location of a ray and its time of arrival. With certain limitations ray theory can also yield transmission loss. Consider Fig. 34.

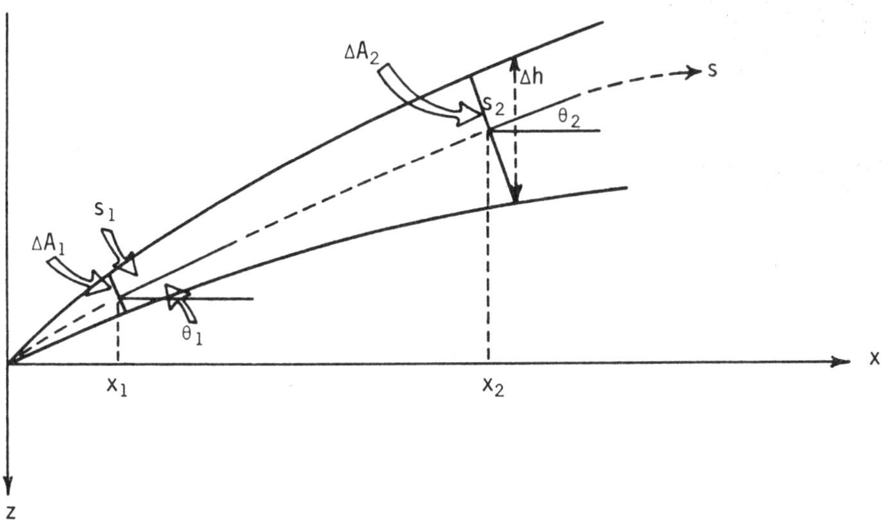

Fig. 34.

The elemental areas, ΔA_1 and ΔA_2, are given by:

$$\Delta A_1 = (s_1 \Delta \theta)(x_1 \Delta \phi_1)$$

$$\Delta A_2 = (\Delta h \cos \theta_2)(x_2 \Delta \phi_2)$$

where ϕ measures angles out of the (x, z)-plane and all other symbols are defined in the above diagram.

If we assume that no energy is absorbed nor leaks out of the propagation tube, the powers at the two points are equal, i.e.:

$$P_1 = P_2$$

Then:

$$I_1 \Delta A_1 = I_2 \Delta A_2$$

which may be solved for the ratio of the intensities to give:

$$\frac{I_1}{I_2} = \frac{\Delta A_2}{\Delta A_1} = \frac{(\Delta h \cos \theta_2)(x_2 \Delta \phi_2)}{(s_1 \Delta \theta)(x_1 \Delta \phi_1)}$$

Assuming:

$$x_1 \simeq s_1 \cos \theta_1 \quad \text{and} \quad \Delta \phi_1 = \Delta \phi_2 = \Delta \phi$$

(the latter is tantamount to assuming that there are no horizontal velocity gradients), the intensity ratio reduces to:

$$\frac{I_1}{I_2} = \frac{x_2 \Delta h c_2}{s_1^2 \Delta \theta c_1}$$

where according to Snell's law:

$$\frac{\cos \theta_2}{\cos \theta_1} = \frac{c_2}{c_1}$$

The transmission loss is defined as:

$$TL = 10 \log I_1/I_2$$

where I_1 is the intensity at one yard. Therefore $s_1 = 1$ and:

$$TL = 10 \log x_2 \Delta h c_2 / \Delta \theta c_1 \tag{6.38}$$

This formula is used to calculate transmission loss from a ray diagram; x_2 is the range of interest, c_1 and c_2 are the sound speeds at the source and the point of interest, respectively, $\Delta\theta$ is the angular separation between chosen rays at the source and Δh is the vertical separation between those rays at the point of interest.

6.2. REFLECTIONS

Let us consider the reflection and transmission of a plane acoustic wave at a plane boundary between two media having different densities and sound speeds, see Fig. 35. Let (x, y, z) be a coordinate system such that the boundary is the $z = 0$ plane, and let the x-axis be parallel to lines of intersection of the wave fronts and the $z = 0$ plane. In this case we may treat the wave in the two coordinates, y and z. Finally we let the plane wave:

$$p_i = A \exp[j(k_i \cdot r - \omega_i t)] \tag{6.39}$$

be incident upon the $z = 0$ plane with an angle θ_i measured from the normal to the plane.

A refelcted wave (p_r) and a transmitted wave (p_2) given by:

$$p_r = B \exp[j(k_r \cdot r - \omega_r t)] \tag{6.40}$$

and:

$$p_2 = C \exp[j(k_2 \cdot r - \omega_2 t)] \tag{6.41}$$

will be generated at the surface, where the coefficients B and C may be complex to account for phase shifts.

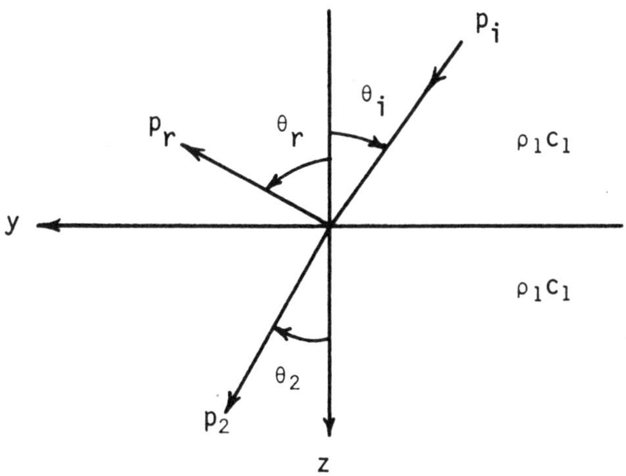

Fig. 35.

The k's are the propagation vectors and the $k \cdot r$'s are:

$$k_i \cdot r = \frac{\omega_i}{c_1}(y \sin\theta_i + z \cos\theta_i)$$

$$k_r \cdot r = \frac{\omega_r}{c_1}(y \sin\theta_r - z \cos\theta_r) \tag{6.42}$$

and:

$$k_2 \cdot r = \frac{\omega_2}{c_2}(y \sin\theta_2 + z \cos\theta_2)$$

Let p_1 and p_2 be the solutions of the wave equation:

$$\frac{\partial^2 p}{\partial z^2} + \frac{\partial^2 p}{\partial y^2} = \frac{1}{c^2}\frac{\partial^2 p}{\partial t^2}$$

in the regions $z \geqslant 0$ and $z \leqslant 0$, respectively. They must satisfy the boundary conditions:

$$p_1(y, 0) = p_2(y, 0)$$

and:

$$\frac{1}{\rho_1}\frac{\partial p_1}{\partial z}\bigg|_{z=0} = \frac{1}{\rho_2}\frac{\partial p_2}{\partial z}\bigg|_{z=0} \tag{6.43}$$

The last requires that the normal particle velocities in the two media be the same at the boundary. The wave p_1 in the upper medium is the sum of the incident and reflected waves, i.e.:

$$p_1 = p_i + p_r \tag{6.44}$$

Since the boundary conditions must apply at $z = 0$ for all y and t, the exponents of p_i, p_r, and p_2 must all be equal. Therefore:

$$\omega_i = \omega_r = \omega_2 = \omega$$

(i.e., there is no frequency change at the boundary), and:

$$\frac{\sin\theta_i}{c_1} = \frac{\sin\theta_r}{c_1} = \frac{\sin\theta_2}{c_2}$$

This latter equation results in Snell's laws, i.e., the law of reflection:

$$\theta_i = \theta_r = \theta_1$$

and the law of refraction:

$$\frac{\sin\theta_1}{c_1} = \frac{\sin\theta_2}{c_2} \tag{6.45}$$

The acoustic waves may now be written:

$$p_i = A\exp\left[j\omega\left(\frac{y\sin\theta_1}{c_1} + \frac{z\cos\theta_1}{c_1} - t\right)\right]$$

$$p_r = B\exp\left[j\omega\left(\frac{y\sin\theta_1}{c_1} - \frac{z\cos\theta_1}{c_1} - t\right)\right]$$

and:

$$p_2 = C\exp\left[j\omega\left(\frac{y\sin\theta_1}{c_1} + \frac{z\cos\theta_2}{c_2} - t\right)\right]$$

We have forced a sinusoidal solution on the field in the lower medium in the vertical direction and, therefore, must be prepared for an unusual treatment of the cosine function.

Applying the boundary condition at $z = 0$, we have:

$$p_i + p_r = p_2 \tag{6.46}$$

and:

$$\frac{\cos\theta_1}{\rho_1 c_1}(p_i - p_r) = \frac{\cos\theta_2}{\rho_2 c_2} p_2$$

Upon dividing both equations of (6.46) by p_i and defining the "reflection coefficient" by:

$$R = p_r/p_i \tag{6.47}$$

and the "transmission coefficient" by:

$$T = p_2/p_i \tag{6.48}$$

we find:

$$T = R + 1$$

and:

$$\frac{\cos\theta_1}{\rho_1 c_1}(1 - R) = \frac{\cos\theta_2}{\rho_2 c_2} T$$

Remember that R and T are in general complex. You may have trouble understanding the physical meaning of these coefficients until you get to the section on intensity.

The above equations may be solved for R and T resulting in:

$$R = \frac{\dfrac{\cos\theta_1}{\rho_1 c_1} - \dfrac{\cos\theta_2}{\rho_2 c_2}}{\dfrac{\cos\theta_1}{\rho_1 c_1} + \dfrac{\cos\theta_2}{\rho_2 c_2}}$$

and:

$$T = \frac{\dfrac{2\cos\theta_1}{\rho_1 c_1}}{\dfrac{\cos\theta_1}{\rho_1 c_1} + \dfrac{\cos\theta_2}{\rho_2 c_2}}$$

Multiplying both equations above by $\rho_2 c_1$ and letting $m = \rho_2/\rho_1$ and $n = c_1/c_2$ (the relative index of refraction for the two media) results in:

$$R = \frac{m\cos\theta_1 - n\cos\theta_2}{m\cos\theta_1 + n\cos\theta_2}$$

and:

$$T = \frac{2m \cos \theta_1}{m \cos \theta_1 + n \cos \theta_2} \tag{6.49}$$

To eliminate θ_2 from T and R we seek Snell's law in a certain form. From equation 6.45 we write:

$$n \sin \theta_2 = \sin \theta_1$$

then:

$$n^2(1 - \cos^2\theta_2) = \sin^2\theta_1$$

and finally:

$$n \cos \theta_2 = \pm (n^2 - \sin^2\theta_1)^{1/2} \tag{6.50}$$

Upon substitution of equation 6.50 into equations 6.49 we obtain:

$$R = \frac{m \cos \theta_1 - \sqrt{n^2 - \sin^2\theta_1}}{m \cos \theta_1 + \sqrt{n^2 - \sin^2\theta_1}}$$

and:

$$T = \frac{2m \cos \theta_1}{m \cos \theta_1 + \sqrt{n^2 - \sin^2\theta_1}} \tag{6.51}$$

where the positive sign of equation 6.50 was taken to insure that $|R| \leqslant 1$ (see next section).

Generally, in acoustics the grazing angle (ϕ) rather than the angle measured from the normal (θ) is used. In this case equations 6.51 become:

$$R = \frac{m \sin \phi - \sqrt{n^2 - \cos^2\phi}}{m \sin \phi + \sqrt{n^2 - \cos^2\phi}}$$

and:

$$T = \frac{2m \sin \phi}{m \sin \phi + \sqrt{n^2 - \cos^2\phi}} \tag{6.52}$$

Intensity

The intensities of the incident, reflected, and transmitted waves are:

$$I_i = \frac{p_i^2}{\rho_1 c_1} = \frac{p_i p_i^*}{\rho_1 c_1}$$

$$I_r = \frac{p_r^2}{\rho_1 c_1} = \frac{p_r p_r^*}{\rho_1 c_1}$$

and: $$I_2 = \frac{p_2^2}{\rho_2 c_2} = \frac{p_2 p_2^*}{\rho_2 c_2}$$

respectively. The intensity ratios are:

$$\frac{I_r}{I_i} = \left(\frac{p_r}{p_i} \cdot \frac{p_r^*}{p_i^*}\right) = RR^* = |R|^2$$

and:

$$\frac{I_2}{I_i} = \frac{\rho_1 c_1}{\rho_2 c_2}\left(\frac{p_2}{p_i} \cdot \frac{p_2^*}{p_i^*}\right) = \frac{n}{m}\, TT^* = \frac{n}{m}\,|T|^2 \qquad (6.53)$$

Reflection loss and transmission loss are given by $10 \log |R|^2$ and $10 \log n/m\,|T|^2$, respectively. Note that $|R| \leqslant 1$ is required but that $|T|$ could be greater than one.

Analysis of reflection at a plane boundary

Due to the many interrelated cases that must be considered, reflection at a boundary is quite complicated. To understand the problem requires a detailed analysis of the various cases. The following analysis is an attempt to be as orderly as possible in presenting the cases. It would probably be worthwhile to refer to the diagrams presented in this section as the section is studied. Reflection at four particular incident angles will be discussed first, then reflection at other angles will be discussed in relation to these.

At normal incidence ($\phi = 90°$) the reflection coefficient is given by:

$$R_n = \frac{m - n}{m + n}$$

If $m \gg n$ (i.e., $\rho_2 c_2 \gg \rho_1 c_1$) then $R_n \simeq 1$ indicating total reflection and zero phase shift. If $m \ll n$ then $R_n \simeq -1$ indicating total reflection but a phase shift of 180°. If $m = n$ (i.e., the acoustic resistances are the same even if the densities and velocities are not) then $R_n = 0$ indicating no reflected wave.

At grazing incidence ($\phi = 0$) the reflection coefficient is given by:

$$R_g = \frac{-\sqrt{n^2 - 1}}{\sqrt{n^2 - 1}} = -1$$

so that reflection is total with a phase shift of 180° for any n and m.

We next define an angle called the "critical angle" by:

$$\cos \phi_0 = n \qquad (6.54)$$

which exists only if $n \leqslant 1$. For incidence at the critical angle the reflection coefficient is given by:

$$R_0 = 1$$

that is, the reflection is total with no phase shift. Physically, the critical angle is the angle of incidence in a low-speed medium for which the transmitted wave propagates parallel to the boundary.

We now define an angle called the "angle of intromission" by:

$$m \sin \phi_I = \sqrt{n^2 - \cos \phi_I}$$

which reduces to:

$$\cos \phi_I = \left(\frac{m^2 - n^2}{m^2 - 1}\right)^{1/2} \tag{6.55}$$

For incidence at the angle of intromission there is no reflected wave, i.e.:

$$R_I = 0$$

In order for an angle of intromission to exist it is necessary that either:

$$m > n > 1 \quad \text{or} \quad m < n < 1$$

For other angles we first consider $n > 1$ and find that R varies from $R_g = -1$ at grazing to $R_n = (m - n)/(m + n)$. If $n > m$ also, R is negative over all angles from grazing to normal. If $n < m$, then an angle of intromission exists where R goes to zero in crossing from negative to positive. Figure 36 illustrates this in terms of the magnitude of R and its phase (ϵ) (for $n > 1$).

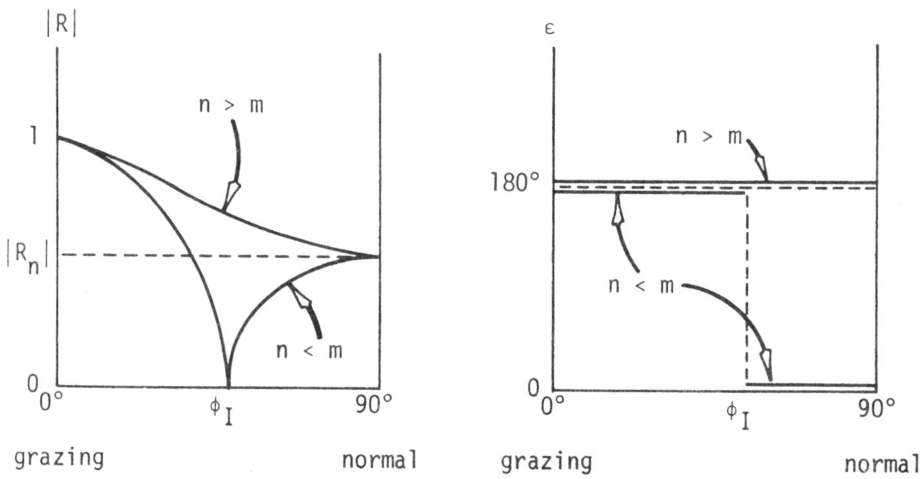

Fig. 36.

Now consider $n < 1$. R varies from $R_g = -1$ at grazing to $R_n = (m - n)/(m + n)$ at normal. A critical angle at which $R_0 = 1$ will always exist in this case.

For angles greater than the critical angle ($\phi > \phi_0$) R is real. If $n < m$, R decreases uniformly from one to a positive R_n. If $n > m$ an angle of intromission, at which $R_I = 0$, also exists (between ϕ_0 and 90°) and R then further decreases to a negative R_n.

For angles less than critical ($\phi < \phi_0$) R is complex and further analysis is required. In this case $\sqrt{n^2 - \cos^2\phi}$ is imaginary so we write it as $j\sqrt{\cos^2\phi - n^2}$ and put this into equation 6.52 to get:

$$R = \frac{m \sin \phi + j\sqrt{\cos^2\phi - n^2}}{m \sin \phi - j\sqrt{\cos^2\phi - n^2}} \tag{6.56}$$

The negative sign of equation 6.50 is to insure that the wave in the second medium decays to zero as z goes to infinity (see next section).

The reflection coefficient has the form:

$$R = \frac{a + jb}{a - jb} = \frac{a^2 - b^2}{a^2 + b^2} + \left(\frac{2ab}{a^2 + b^2}\right) j$$

The magnitude of R is unity for all $\phi < \phi_0$ because:

$$|R| = \sqrt{RR^*} = \left(\frac{a + jb}{a - jb} \cdot \frac{a - jb}{a + jb}\right)^{1/2} = 1$$

The reflection coefficient may be written in the form:

$$R = |R| \, e^{j\epsilon} = e^{j\epsilon}$$

where the phase is given by:

$$\tan \epsilon = \frac{2ab}{a^2 - b^2} = \frac{2b/a}{1 - \dfrac{b^2}{a^2}}$$

If we define $\tan \delta = b/a$ this reduces to:

$$\tan \epsilon = \frac{2 \tan \delta}{1 - \tan^2\delta} = \tan 2\delta$$

where the latter part of the equation is completed by a trigonometric relation. Therefore, $2\delta = \epsilon$ and:

$$\tan \frac{\epsilon}{2} = \frac{b}{a} = \frac{\sqrt{\cos^2\phi - n^2}}{m \sin \phi} \tag{6.57}$$

At critical incidence $\epsilon = 0°$ and at grazing $\epsilon = 180°$. For angles between equation 6.57 gives a uniform curve.

Figure 37 illustrates the above discussion of the reflection coefficient for $n < 1$ in terms of $|R|$ and ϵ.

Solution in the lower medium for the case of total reflection at angles below critical

The solution in the lower medium is:

$$p_2 = C \exp \left[j\omega \left(y \frac{\cos \phi}{c_1} + z \frac{\sin \phi_2}{c_2} - t \right) \right]$$

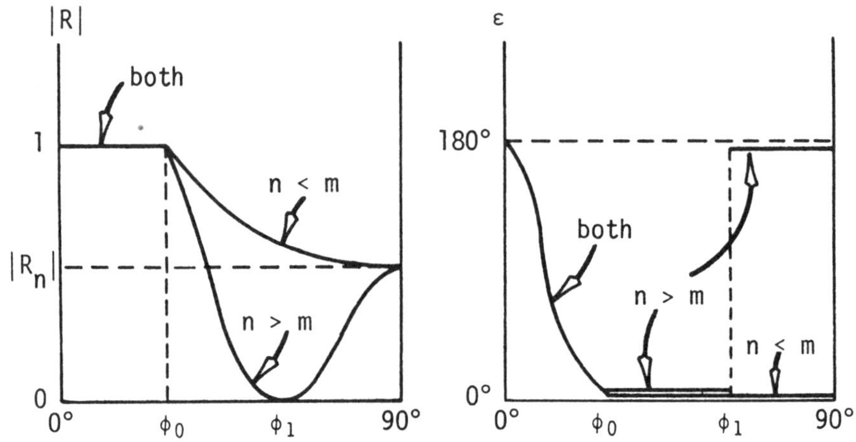

Fig. 37.

But for incident angles less than critical ($\phi < \phi_0$), $\sin \phi_2$ is complex and is given by:

$$\sin \phi_2 = +\frac{j}{n}\sqrt{\cos^2\phi - n^2}$$

(The positive sign was chosen to insure that the wave in the second medium dies away as z goes to infinity.) The solution is now:

$$p_2 = C \exp\left[-\frac{z\omega}{c_2 n}\sqrt{\cos^2\phi - n^2} + j\omega\,\frac{\cos\phi}{c_1}\left(y - \frac{c_1}{\cos\phi}\,t\right)\right]$$

or:

$$p_2 = C \exp\left[-\frac{z\omega}{c_1}\sqrt{\cos^2\phi - n^2} + j\,\frac{\omega}{c'}\,(y - c't)\right] \qquad (6.58)$$

Note that the wave propagates along the boundary at the velocity $c' = c_1/\cos\phi$ and is attenuated in the lower medium (where z is positive) in accordance with the real part of the exponent. These points are illustrated in Fig. 38.

These results suggest that a constant acoustical energy is set up in the lower medium and that this energy is needed to produce the total reflection. In the steady state this energy does not change in time and no power flows cross the boundary into an ideal lower medium. But if the lower medium is lossy, energy is dissipated in it and a continuous power flow across the boundary is needed to replace the lost energy, thus giving rise to a reflection that is not quite total.

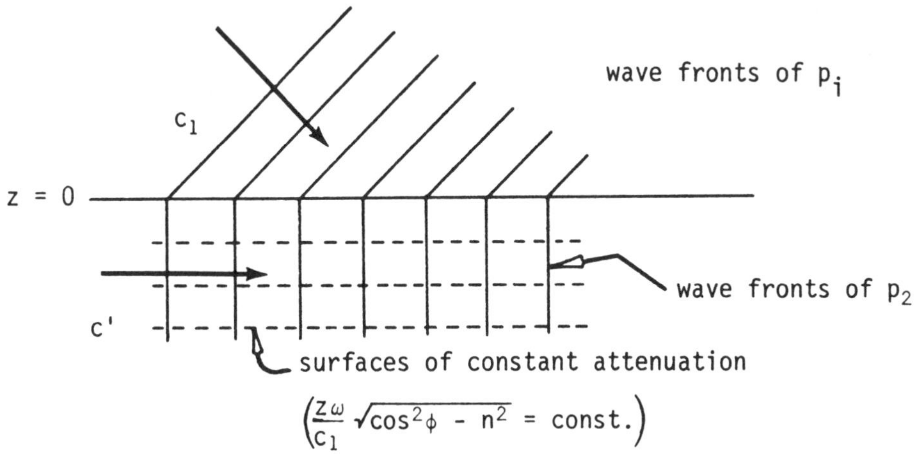

Fig. 38.

Reflections at the air—water interface

Let us consider the two cases of reflections at the air—water boundary as illustrated in Fig. 39.

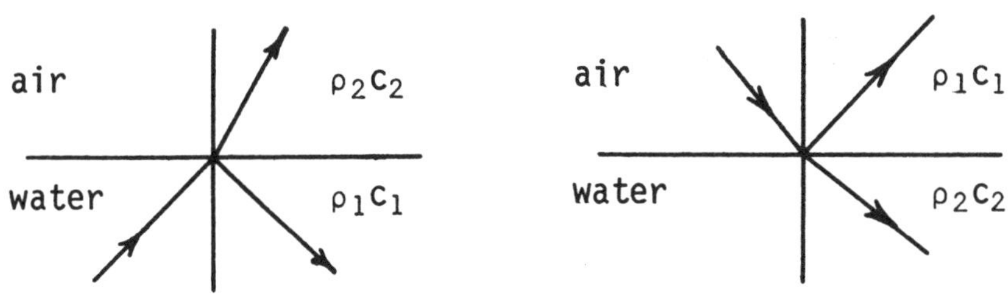

Fig. 39.

Case A

$$n = \frac{c_1}{c_2} = 4.3$$

$$m = \frac{\rho_2}{\rho_1} = \frac{1}{770}$$

$$n > 1, \quad m < 1$$

Case B

$$n = \frac{c_1}{c_2} = \frac{1}{4.3}$$

$$m = \frac{\rho_2}{\rho_1} = 770$$

$$n < 1, \quad m > 1$$

In Case A:

$$R = \frac{\frac{1}{770}\sin\phi - \sqrt{4.3^2 - \cos^2\phi}}{\frac{1}{770}\sin\phi + \sqrt{4.3^2 - \cos^2\phi}}$$

Over the full range of ϕ, R is very nearly -1. So the wave is almost totally reflected and is 180° out of phase with the incident wave. This is as we would expect because an air boundary over water may be considered as a free boundary that will not support pressure, i.e.:

$$p_1 = p_i + p_r = p_2 = 0 \quad \text{or} \quad p_i = -p_r$$

at $z = 0$. So that $R = -1$. This is generally the boundary condition imposed at the sea surface.

The transmission loss across the boundary may be determined by:

$$TL = -10 \log \frac{n}{m} |T|^2$$

For normal incidence this is:

$$TL_n = -10 \log \frac{n}{m} |T_n|^2 = -10 \log \frac{4nm}{(m+n)^2}$$

In our case:

$$TL_n = -10 \log \left(\frac{4}{770 \times 4.3}\right) = 29 \text{ dB}$$

In Case B:

$$R = \frac{770 \sin \phi - \sqrt{\left(\frac{1}{4.3}\right)^2 - \cos^2\phi}}{770 \sin \phi + \sqrt{\left(\frac{1}{4.3}\right)^2 - \cos^2\phi}}$$

In this case a critical angle exists. It is given by:

$$\cos \phi_0 = n = \frac{1}{4.3} \quad \text{or} \quad \phi_0 = 77°$$

Below 77° the magnitude of the reflection coefficient is unity, but there is a non-zero phase shift. Above 77° the coefficient is still very nearly unity and there is no phase change.

The transmission loss at normal incidence is:

$$TL_n = -10 \log \left[\frac{4nm}{(m+n)^2}\right] = 29 \text{ dB}$$

Bottom reflections

The most common condition at the bottom boundary is $n < 1$ and $m > 1$. In this case a critical angle exists and, assuming a lossless bottom, total reflection occurs for angles more grazing than the critical angle. Actually bottoms are not lossless and total reflection does not occur at and just below the critical angle.

For some soft mud bottoms n is greater than one, while m is still greater than one. If the case $m > n > 1$ holds an angle of intromission, for which transmission is total, exists.

Interference between reflected and direct rays: Lloyd's mirror effect

Consider the problem of interference between a reflected and direct ray. Let the surface have a reflection coefficient R, which may be complex to account for phase shifts:

$$R = |R| e^{j\epsilon} \tag{6.59}$$

Let p' be the pressure at a point $P(x, z)$ due to the direct ray from an object at 0, and let p'' be the pressure at the point P due to the image at I, see Fig. 40.

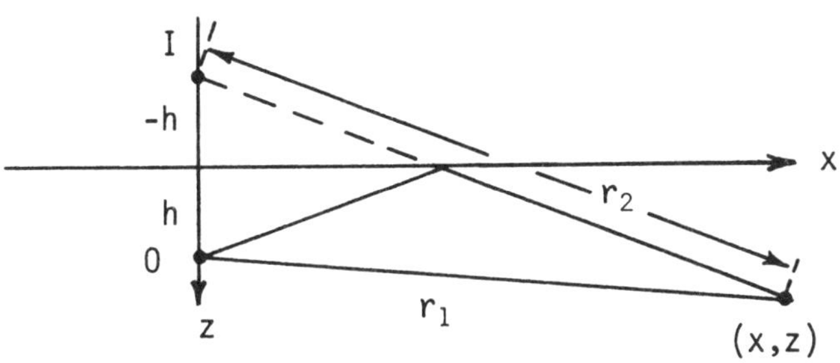

Fig. 40.

Then:

$$p' = \frac{A}{r_1} e^{j\omega t}$$

where A is assumed to be real, and:

$$p'' = \frac{RA}{r_2} e^{[j\omega(t+\tau)]}$$

where $\omega\tau$ is the phase shift due to the difference $(r_2 - r_1)$ in propagation distance:

$$\tau = (r_2 - r_1)/c$$

The pressure at P is:

$$p(x, z, t) = p' + p'' = A e^{j\omega t}\left(\frac{1}{r_1} + \frac{R}{r_2} e^{j\omega\tau}\right)$$

Now:

$$r_1 = \sqrt{x^2 + (z-h)^2}$$

and:

$$r_2 = \sqrt{x^2 + (z+h)^2}$$

Let us approximate the radicals by:

$$r_1 \simeq x \left[1 + \frac{1}{2}\frac{(z-h)^2}{x^2}\right]$$

and:

$$r_2 \simeq x \left[1 + \frac{1}{2}\frac{(z+h)^2}{x^2}\right]$$

which is good for $x \gg z, h$. Then the difference is:

$$r_2 - r_1 \simeq 2zh/x$$

Now, except for the difference noted above, it is assumed that x is so large that we may take:

$$r_1 = r_2 = x$$

We now write the pressure:

$$p(x, z, t) = \frac{A\,e^{j\omega t}}{x}\left[1 + R\exp\left\{j\left(\frac{2\omega zh}{cx}\right)\right\}\right]$$

In terms of trigonometric functions we have:

$$p(x, z, t) = \frac{A\,e^{j\omega t}}{x}\,[1 + R\cos a + jR\sin a]$$

where:

$$a = \frac{2\omega zh}{cx} = \frac{4\pi zh}{x\lambda} \tag{6.60}$$

The intensity at the point P is given by:

$$I(x, z) = pp^*/\rho c$$

Then:

$$I(x, z) = \frac{A^2}{\rho c x^2}[1 + (R + R^*)\cos a + j(R - R^*)\sin a + RR^*\cos^2 a$$
$$+ RR^*\sin^2 a]$$

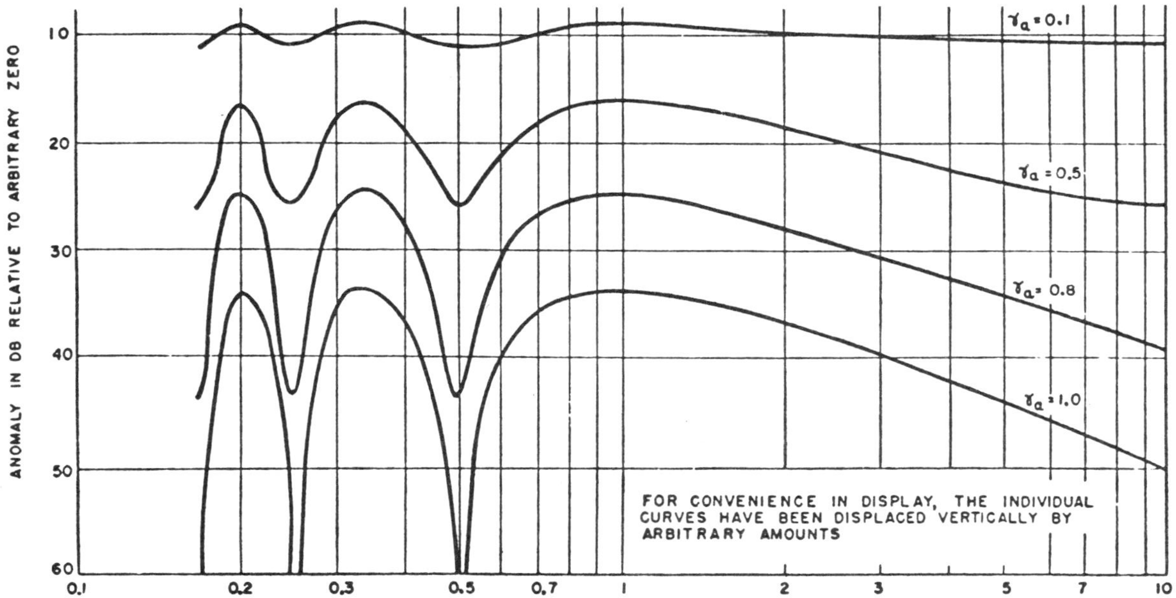
ANOMALY IN DB RELATIVE TO ARBITRARY ZERO
10
20
30
40
50
60
0.1
0.2
0.3
0.5
0.7
1
2
3
5
7
10
$\gamma_a = 0.1$
$\gamma_a = 0.5$
$\gamma_a = 0.8$
$\gamma_a = 1.0$
FOR CONVENIENCE IN DISPLAY, THE INDIVIDUAL CURVES HAVE BEEN DISPLACED VERTICALLY BY ARBITRARY AMOUNTS

Fig. 41.

or:

$$I(x, z) = \frac{A^2}{\rho c x^2}[1 + 2|R| \cos \epsilon \cos a + 2|R| \sin \epsilon \sin a + |R|^2]$$

Finally this may be written:

$$I(x, z) = \frac{A^2}{\rho c x^2}[1 + 2|R| \cos (a - \epsilon) + |R|^2] \tag{6.61}$$

If the phase shifts upon reflection by 180° (situation for a water-to-air interface), equation 6.61 reduces to:

$$I(x, z) = \frac{A^2}{\rho c x^2}[1 - 2|R| \cos a + |R|^2] \tag{6.62}$$

where a is given by equation 6.60. Plotted in Fig. 41* in arbitrary units versus the dimensionless range $\lambda x/4zh$ is $I(x, z)x^2$.

If, in addition to the 180° phase shift, $|R| = 1$ (which is a good approximation for the water-to-air interface), then (6.62) reduces to:

$$I(x, z) = \frac{2A^2}{\rho c x^2}(1 - \cos a) \tag{6.63}$$

For large enough x we can expand the cosine term into a series and retain only the first two terms, i.e.:

$$\cos a \simeq 1 - a^2/2$$

Then (6.63) becomes:

$$I(x, z) = \frac{A^2 a^2}{\rho c x^2} = \frac{16\pi^2 z^2 h^2 A^2}{\rho c \lambda^2} x^{-4} \tag{6.64}$$

The inverse fourth power range dependent acoustic field described by this equation is sometimes called a "dipole field".

6.3. THE THEORY OF SHALLOW-WATER ACOUSTIC PROPAGATION

Basically, there are two methods of developing the theory of shallow-water acoustic propagation: the method of images and of normal modes. Actually, as we shall do here, one can develop a solution for the shallow-water problem in terms of a sum of images then transform this series into a sum of normal modes (i.e., the normal-mode solution).

According to this approach, one attempts to construct a solution to:

$$\nabla^2\phi + k^2\phi = 0$$

---

* From *Physics of Sound in the Sea*, Natl. Defense Res. Comm. Natl. Res. Council Div. 6 Sum. Tech. Rept. 8, Chap. 4, fig. 10, 1946.

where ϕ is the velocity potential or the pressure and $k = \omega/c$, and a set of boundary conditions by summing over an infinite set of images. If we may consider the boundaries to be a perfectly free surface at the top and a perfectly rigid surface at the bottom, the conditions are:

$$\partial\phi/\partial z = 0 \text{ at } z = 0 \quad \text{and} \quad \phi = 0 \text{ at } z = h$$

The problem is illustrated in Fig. 42.

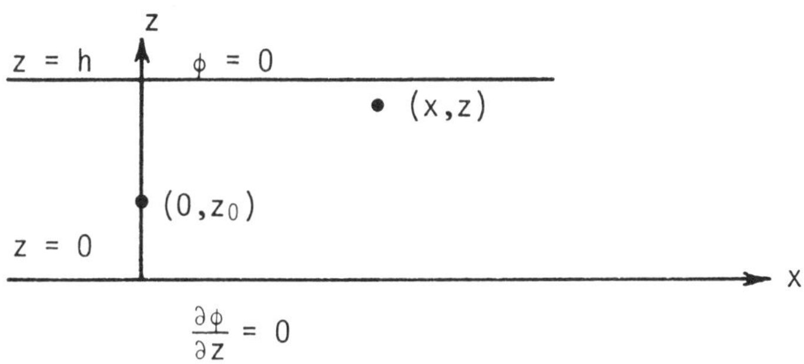

Fig. 42.

The source is at $(0, z_0)$ and the point of interest is (x, z).

The solution is formed by the following sum over the images:

$$\phi = \phi_0 \sum_{m=0}^{\infty} R_m \frac{\exp(jkr_m)}{r_m} \tag{6.65}$$

where r_m is the distance to the mth image and R_m is the mth order reflection coefficient given by:

$$R_m = \prod_{i=0}^{m} R_i \tag{6.66}$$

where R_i is the reflection coefficient for a single reflection. In our case R_i is plus or minus one depending upon whether the reflection occurred at the bottom or the top, respectively.

By making an integral expansion of $\exp(jkr_m)/r_m$* the series image solution may be transformed into a normal-mode solution (a sum over normal modes instead of over images). The solution is given by**:

$$\phi(x, z) = \frac{2\pi j\phi_0}{h} \sum_{l=0}^{\infty} \cosh b_l z_0 \cosh b_l z H_0^{(1)}(\xi_l x) \tag{6.67}$$

---

* Brekhovskikh, L.M., *Waves in Layered Media*, p. 332. Academic Press, New York, 1960.
** Adapted from Brekhovskikh, *ibid.*, p. 338.

where:

$$b_l = \frac{j(l+\frac{1}{2})\pi}{h} \quad \text{and} \quad \xi_l = \sqrt{b_l^2 + k^2}$$

and $H_0^{(1)}(\xi_l x)$ is the Hankel function of the first kind. A plot of this function is not markedly different from Fig. 31 (with z being the vertical axis).

6.4. SPREADING AND ATTENUATION

Spreading

We have already briefly discussed spherical and cylindrical spreading. In general, we may write the spreading law as follows:

$$TL_{\text{spread}} = n\,(10 \log r) \qquad (6.68)$$

where r is in yards and n depends on the type of spreading, i.e.:

$n = 0$ no spreading
$n = 1$ cylindrical spreading
$n = 2$ spherical spreading
$n = 3$ spherical spreading with linear time stretching
$n = 4$ "dipole" type spreading associated with the Lloyd's mirror effect.

The case of no spreading is of no significance in practical acoustics. Cylindrical spreading may apply in the case of shallow water, if the boundaries are not too lossy, or in the case of surface channel or deep channel propagation.

Spherical spreading generally is good at an intermediate range far enough from the source to assure far field but near enough so that we may assume free field propagation (i.e., closer than the nearest boundary or gross medium change).

Time spreading in a free field results in $n = 3$ but is generally a hypothetical case. But one can imagine a situation where, due to multipath propagation, the energy of a pulse becomes spread out over time as well as space. This is generally what one attempts to account for when energy flux density is used rather than intensity.

In general, the spreading law for sound propagation in the sea is not simple, not only because of the reflection at the boundaries, but also because of the refraction that takes place due to sound-speed gradients.

Attenuation

Attenuation of sound is divided into two classes, scattering and absorption. Scattering results from the reflection, and the resonant absorption and re-radiation diffraction, of sound by macroscopic and microscopic inhomogeneities in the medium. Absorption is due to the following phenomena: (1) thermal conductivity; (2) viscosity; (3) structural and chemical relaxations; and (4) resonant absorption.

Scattering and absorption are described by the same mathematical formula and will in general be called attenuation. The loss due to spreading follows a different formula and is not included in attenuation.

The fractional infinitesimal change in intensity ($\mathrm{d}I/I$) is proportional to the infinitesimal distance travelled ($\mathrm{d}r$), i.e.:

$$\frac{\mathrm{d}I}{I} = -2\delta\,\mathrm{d}r$$

where δ is a positive constant called the "(pressure) attenuation coefficient" in neper/meter and the minus sign indicates that the change is a loss. Integrating this equation between r and r_0 results in:

$$\ln I - \ln I_0 = -2\delta(r - r_0)$$

or:

$$I = I_0 \exp\left[-2\delta(r - r_0)\right] \qquad (6.69)$$

It is more common to write equation 6.69 as:

$$I = I_0\, 10^{-(\alpha/10)(r-r_0)} \qquad (6.70)$$

where r is measured in kyd and α is the (intensity) attenuation coefficient in dB/kyd.

A relation between the two attenuation coefficients may be obtained by equating the exponentials of (6.69) and (6.70), i.e.:

$$10^{-(\alpha/10)(r-r_0)} = \exp\left[-2\delta(r - r_0)\right]$$

where r on the left is in kyd and on the right in meters. This reduces to:

$$\alpha = \left[20 \log e \left(\frac{\mathrm{dB}}{\mathrm{neper}}\right)\right] \left[1.1 \cdot 10^3 \left(\frac{\mathrm{meter}}{\mathrm{kyd}}\right)\right] \delta$$

$$\alpha = 9.6 \cdot 10^3\, \delta \left(\frac{\mathrm{dB}}{\mathrm{kyd}}\right) \qquad (6.71)$$

The transmission loss due to attenuation is given by:

$$TL_{\mathrm{Att}} = -10 \log I/I_0 = \alpha(r - r_0) \qquad (6.72)$$

r_0 is taken to be one yard (0.001 kyd) and I_0 is the source intensity. Generally r_0 may be neglected as compared to r, so that the transmission loss due to attenuation may be written:

$$TL_{\mathrm{Att}} = \alpha r \qquad (6.73)$$

The total transmission loss may be written:

$$TL = \alpha r + 10\, n \log r + 30\, n \tag{6.74}$$

where r is in kyd and the term $30n$ accounts for the conversion from yards to kyd, and n is determined by the type of spreading.

PROBLEMS

1. Using the integral equation (equation 6.28) in unbounded space containing the point source:

   $$\psi(r') = a\delta(r' - r_0)$$

   and the free space Green's function:

   $$G(r, r') = \frac{\exp(jk|r - r'|)}{|r - r'|}$$

   find an expression for the pressure field $p(r)$.
2. Using the integral equation find the pressure field in the half-space $z > 0$ if there is a source $\psi(r') = a\delta(r' - r_0)$ at the point $r_0 = (0, 0, z_0)$ and an infinite plane S at $z = 0$ on which $p = 0$. Hint: construct a Green's function from point sources such that $G = 0$ on S.

CHAPTER 7

SOUND PROPAGATION IN THE SEA

Actual sound propagation in the sea is complicated by many factors: physical and chemical properties of sea water cause attenuation and refraction; rough and poorly defined surfaces complicate reflection; and ambient noise and reverberation present problems in detection. In this chapter we shall consider refraction due to sound-speed gradients, channeling due to refraction and reflection, and attenuation due to various causes. Problems of ambient noise and reverberation will be considered in latter chapters.

7.1. SOUND-SPEED PROFILES IN THE SEA

The speed of sound in water is approximately 1500 m/sec, but its precise value is strongly dependent upon temperature, pressure, and to a lesser extent, upon salinity. Generally, it increases as each of these quantities increases.

What is perhaps the most accurate empirical formula was provided by Wilson.* This formula is:

$$c = 1{,}449.14 + V_{\mathrm{T}} + V_{\mathrm{P}} + V_{\mathrm{S}} + V_{\mathrm{STP}} \tag{7.1}$$

where V_{T}, V_{P}, and V_{S} are fourth, fourth, and second-order polynomials in the temperature, pressure, and salinity, respectively. V_{STP} is a polynomial involving cross-products of S, T, and P. But this formula is probably too complicated for general use. One can truncate the formula into as many terms as one desires, but this produces less accurate results than to use a polynomial obtained from a least-squares fit with the desired number of terms.

Using equation 7.1 and the most accurately determined temperature, salinity, and pressure (i.e., as obtained from oceanographic data), the speed of sound can be calculated with an accuracy of about ± 0.3 m/sec. Generally, the speed of sound is determined in the upper few hundred meters using typical values for salinity and bathythermograph (BT) data for temperature. These results are less accurate but the data are much easier to obtain.

A velocity of sound measuring device called the "sing-around velocimeter" is coming into quite common use. Presently, it provides results with the

* Wilson, W.D., J. Acoust. Soc. Am., **23**: 1357 (1960).

same accuracy as the oceanographic data but is easier to obtain and provides continuous (or quasi-continuous) results with depth.

The instrument operates on the simple principle of timing the travel of an acoustic pulse between two points, but the method of timing provides the "sing-around" aspect. Each time an acoustic pulse is received by the receiver, electronic circuits trigger another acoustic pulse. The frequency or period of the generated pulses is measured and related to the travel time. Future generations of these devices are expected to improve the accuracy by an order of magnitude.

Typical vertical variations in temperature and salinity are 25°C and 2 ppt, respectively. These variations produce a sound-speed variation of about 80 and 3 m/sec, respectively. The sound-speed variation due to pressure alone between the surface and 3,000 m is about 50 m/sec. Thus, we see that temperature and pressure are the important variables in determining the vertical profile of the sound speed. Pressure varies linearly with depth to a very close approximation; therefore, temperature is the primary variable that must be measured at sea in order to determine the sound speed. For this purpose BT data are of special interest.

Figure 43 indicates typical temperature profiles and their general latitude dependence.

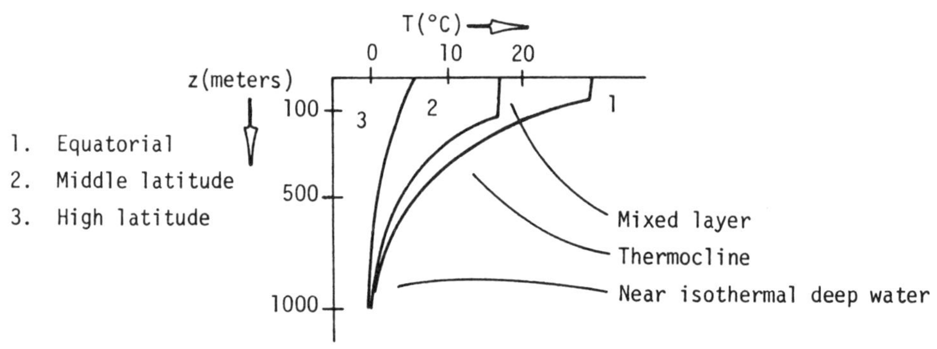

Fig. 43.

The sound-speed profiles are determined primarily by the temperature and pressure profiles. Figure 44 shows sound-speed profiles for the typical pressure and temperature profiles at the right.

Since gradients of sound speed cause refraction, sound-speed profiles are extremely important in practical acoustic propagation problems. It is convenient to remember that acoustic rays are bent (refracted) toward regions of lower speed. For this reason the sound-speed minima are very important.

Figure 45 shows how a typical sound-speed profile determines the path of a ray. In addition to showing the refracted paths of the rays, this figure shows regions called "shadow zones" where direct rays cannot enter.

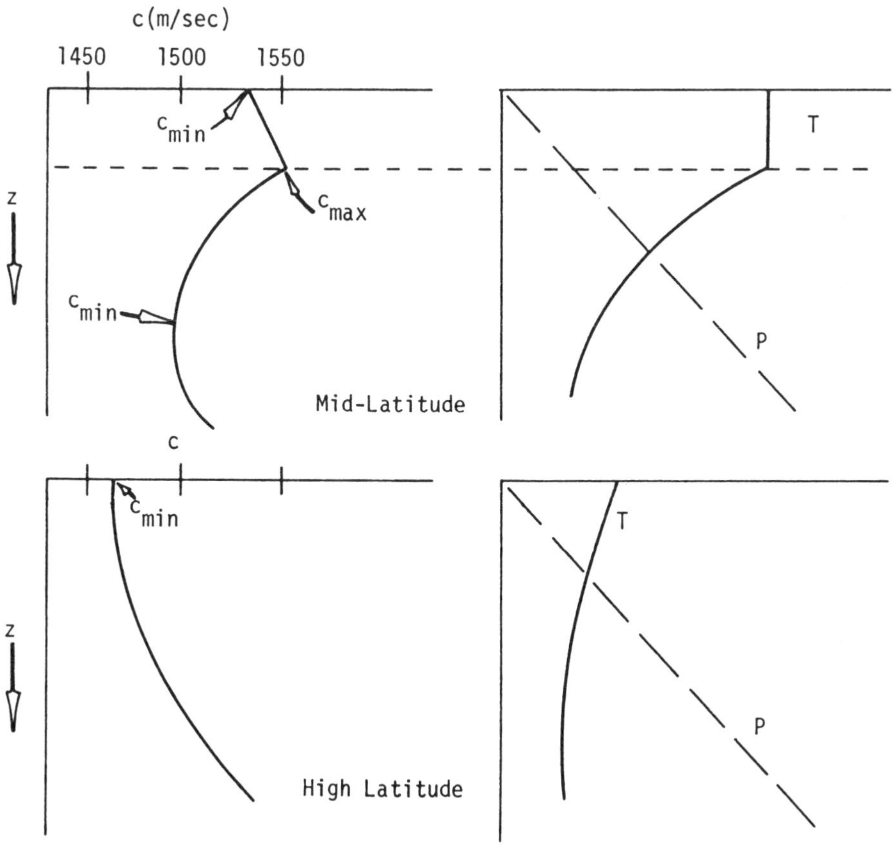
c(m/sec)
1450
1500
1550
c_{min}
c_{max}
z
c_{min}
T
P
Mid-Latitude
c
c_{min}
z
T
P
High Latitude

Fig. 44.

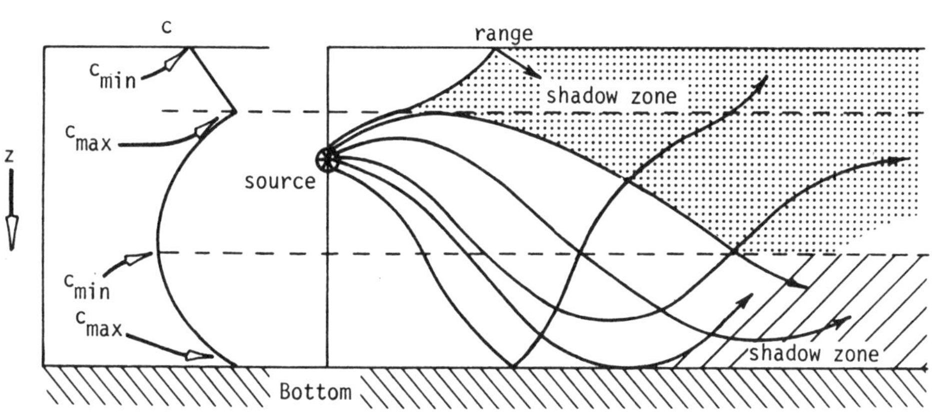
c
c_{min}
c_{max}
z
c_{min}
c_{max}
range
shadow zone
source
shadow zone
Bottom

Fig. 45.

Shadow zones begin at sound-speed maxima whether the maxima occurs in the column or at a boundary.

7.2. SOUND CHANNELING

Mixed-layer sound channel

Any horizontal layer of the ocean is potentially a sound channel; i.e., a channel or duct, which, due to refraction, produces a spreading less than spherical. Often, an isothermal surface layer forms such a sound channel. Due to the pressure effect the sound speed in this layer increases from the surface to the top of the thermocline.

Figure 46 shows acoustic ray paths for such a channel.

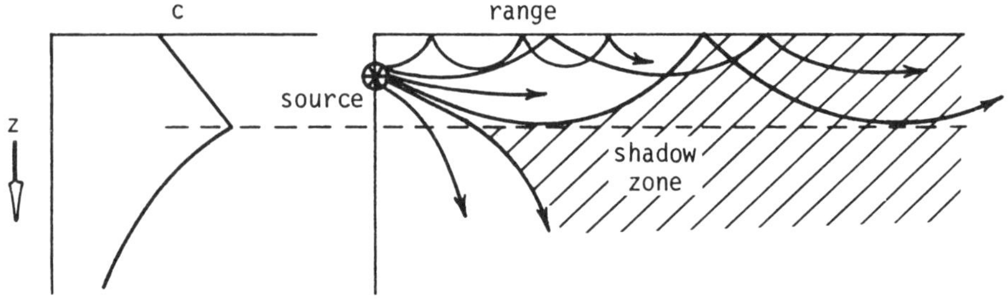

Fig. 46.

The existence of the mixed-layer sound channel is important to the operation of hull-mounted surface-ship sonars.

Due to scattering at the surface and "leakage" out the bottom, spreading in this sound channel is, generally, not much better than spherical. A "leakage coefficient" (α_L) that represents an energy loss in addition to cylindrical-type spreading may be defined. This coefficient is a function of surface roughness, mixed-layer thickness, sound-speed gradient below the layer, and frequency (leakage at low frequency is greater than it is at high frequency).

A simple model for transmission loss in a channel has been given by:

$$TL = 10 \log r + 10 \log r_t + (\alpha + \alpha_L)\, r \cdot 10^{-3} \qquad (7.2)$$

where range is in yards and r_t is a transition range to which the spreading is spherical and for $r \gg r_t$ the spreading is cylindrical. r_t is somewhat arbitrary but has a value of a few kyd.

A semi-quantitative plot of transmission gain over spherical transmission loss is plotted in Fig. 47 for a moderate range of about 15 kyd.

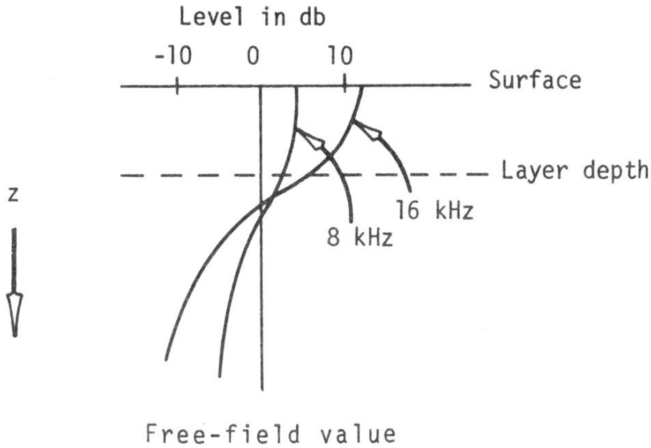

Fig. 47.

At very low frequency, sound ceases to be trapped in the layer simply because the wavelength becomes too large to be contained in the layer. The maximum wavelength (λ_{max}) that can be trapped is given by:

$$\lambda_{max} = 4.7 \cdot 10^{-3} H^{3/2} \tag{7.3}$$

where H is the mixed layer depth and both λ_{max} and H are in feet. For example, for a mixed layer depth of 100 feet, λ_{max} is 4.7 feet, which corresponds to a frequency of 1.1 kHz.

The deep sound channel

The sound-speed minimum that occurs at great depth as seen in previous diagrams is the axis of another sound channel called the "deep sound channel" or "*SOFAR*" (Sound Fixing And Ranging). This minimum is produced by the combined effects of decreasing temperature and increasing pressure and occurs at depths of 800—1200 m in low and mid-latitudes but approaches the surface in high latitudes.

This sound channel was originally investigated by Ewing and Worzel* in 1948. Figure 48 was taken from that reference. It indicates ray paths for propagation in the deep sound channel for a source on the axis.

Transmission in the deep sound channel is exceptionally good because the rays are channeled primarily by refraction rather than reflection (associated with reflections are generally quite severe energy losses). Sound transmission in the deep sound channel is measured in thousands of miles.

Due to multipath transmission, this long-range propagation is characterized by severe distortions in the signal. A pulse of a few tenths of milliseconds,

---

* From Ewing, M. and Worzel, J.L., *Long-range sound transmission*. Geol. Soc. Am. Mem. 27, 1948. Used with permission of the Geological Society of America.

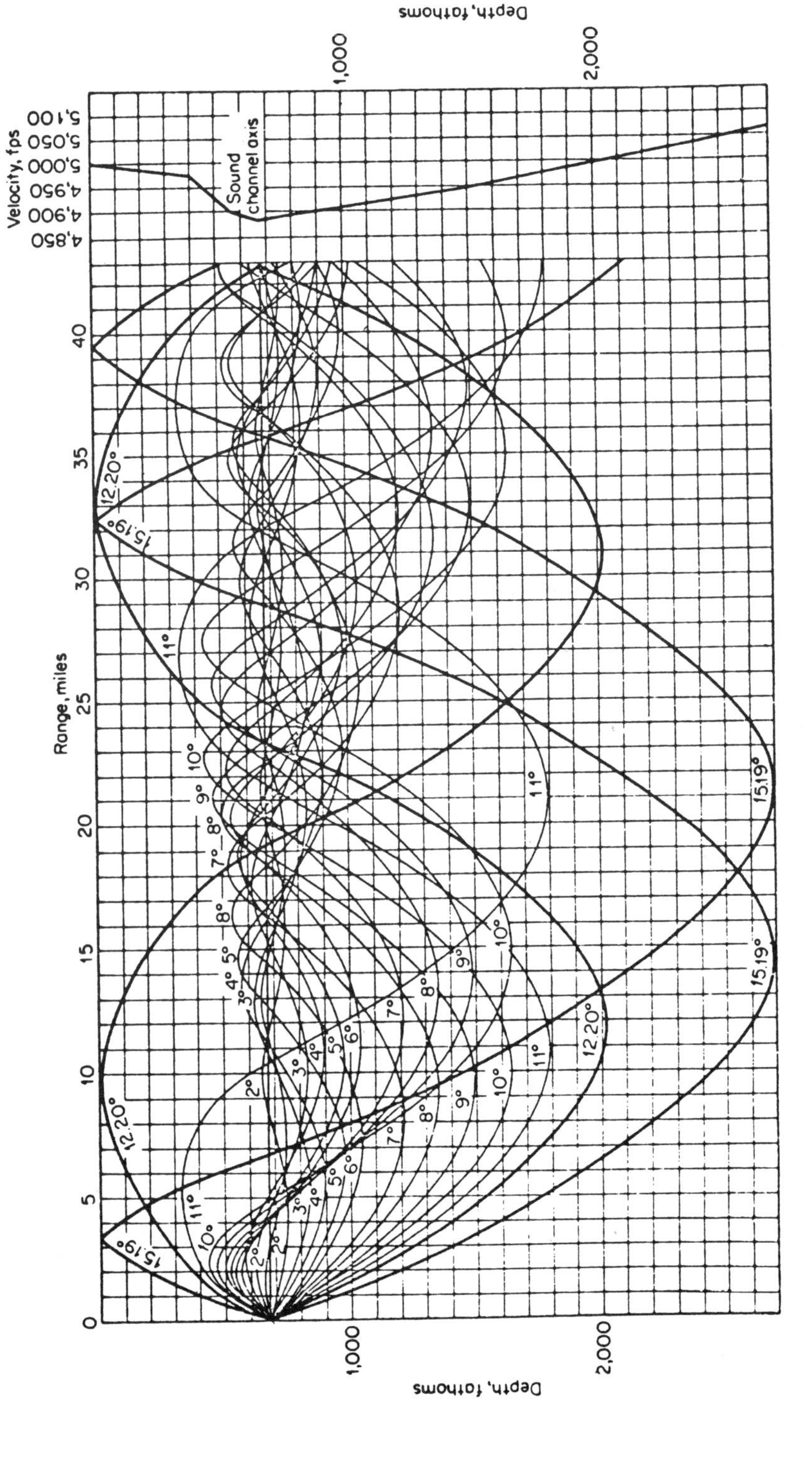

Depth, fathoms
1,000
2,000
Velocity, fps
4,850
4,900
4,950
5,000
5,050
5,100
Sound channel axis
Range, miles
0
5
10
15
20
25
30
35
40
15.19°
12.20°
11°
10°
9°
8°
7°
6°
5°
4°
3°
2°
Depth, fathoms
1,000
2,000

Fig. 48.

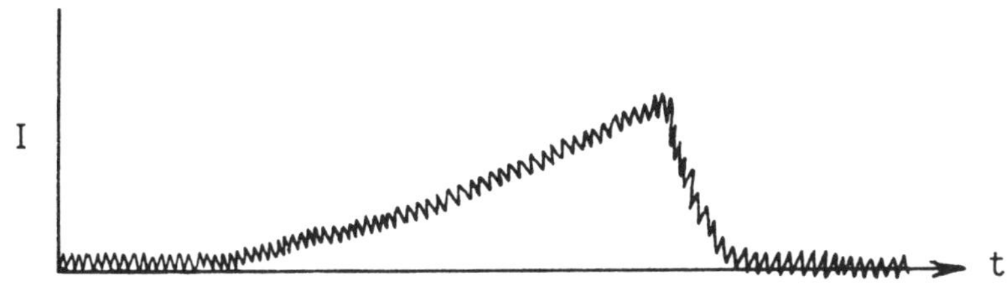

Fig. 49.

such as that of an explosion, becomes stretched out at the rate of about 9.4 sec per 1000 miles of travel. The received signal is characterized by a slow build-up of intensity followed by a sharp drop, such as is indicated in Fig. 49. The initial rise in intensity is produced by the arrival of rays making wide excursions from the sound-speed minimum. These rays travel the fastest because, although their paths are slightly longer, they spend most of their time in a higher velocity region. But these rays carry the least energy. The energy at the end of the pulse is due to the bundle of rays travelling near the acoustic axis. Once these rays pass the signal ceases.

An important feature related to the propagation of sound in this channel is the existence of regions containing a number of caustics. These regions are called "convergence zones" and intensity of sound in these regions is generally 10—20 dB higher than the free field value. Figure 50 illustrates the character of convergence zones for a shallow source and receiver.

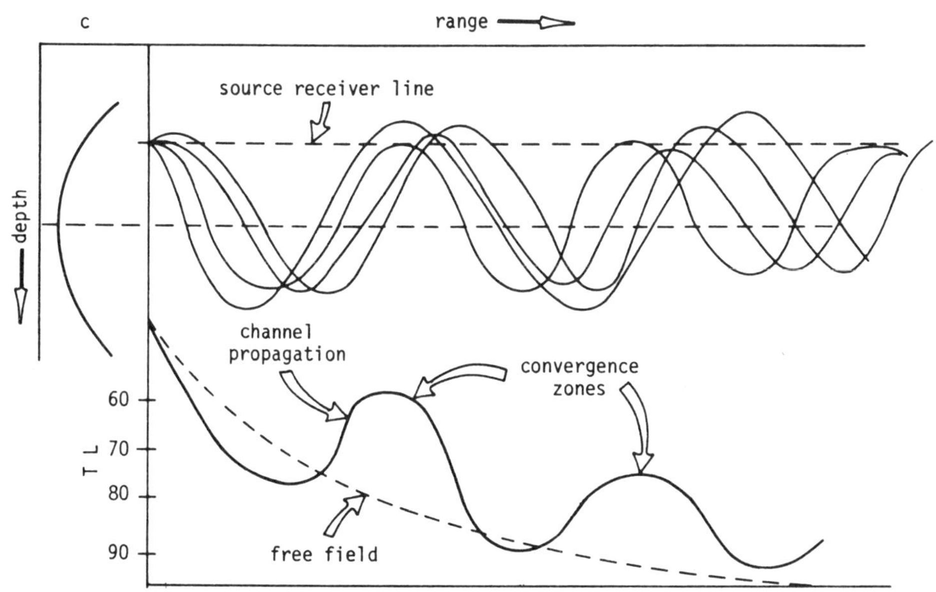

Fig. 50.

The increase in intensities associated with the peak in the transmission loss diagrams are called "convergence gains". These convergence zones occur in approximately 30 mile intervals and are about 3 miles wide.

Shallow-water transmission

Shallow-water sound propagation is strongly dependent upon the depth and sound-speed profile. If the profile has a negative gradient from the top to the bottom, propagation is particularly dependent upon the bottom configuration and bottom and sub-bottom properties. This is because rays are refracted toward the bottom. If the profile has a positive gradient, surface roughness becomes the important factor.

Propagation in shallow water may be described by either ray theory or normal-mode theory depending upon the ratio of wavelength (λ) to depth (d). If:

$$\frac{\lambda}{d} = \left(\frac{c}{f_L d}\right) < \frac{1}{10}$$

ray theory is generally more appropriate. For example, the lower frequency limit for the applicability of ray theory in water of 15 m depth is:

$$f_L = \frac{10c}{d} = \frac{15{,}000}{d} = \frac{15{,}000}{15} = 1000\,\text{Hz}$$

Normal-mode solutions are better suited if:

$$\lambda/d > 1/2$$

For example, in the above case the upper frequency limit for a good normal-mode analysis is:

$$f_u = \frac{2c}{d} = \frac{3{,}000}{15} = 200\,\text{Hz}$$

Normal-mode solutions apply for higher frequencies but are less practical because of the number of terms needed in the series solution (see equation 6.67).

Let us consider the ray solution. Over the range of validity for the ray solution, the reflection coefficient is given by equation 6.52. It is generally assumed that the lower boundary condition is characterized by a sound-speed and medium density increase on going from the water into the sediments. In this case a critical angle exists so that the reflection coefficient is unity for angles from grazing up to the critical angle and the phase shift decreases in a regular manner from 180° at grazing to 0° at the critical angle. From the critical angle up to normal incidence, the reflectivity decreases from one to a finite value, while the phase shift remains zero. The condition at the surface is described by the free-surface boundary condition for which reflection is total and the phase shift is 180° at all angles.

The basic form of the propagation depends upon the sound-speed structure, while the details of the propagation depend upon the specific environmental properties as discussed before. Let us consider three specific profiles: the isovelocity profile, the negative-gradient profile, and the positive-gradient profile. Combinations are also found but the analysis of the propagation provides nothing essentially different from what was discussed earlier and what will be discussed now.

Let us first consider the isovelocity profile and, for the moment, assume that the source and receiver are at the surface. Then the only effect is straight line propagation with reflections from the surface and the bottom.

Let *Bn* represent the *n*th-order bottom-reflected ray, i.e., the ray that has been reflected from the bottom *n* times (see Fig. 51). You will note that the *n*th-order bottom-reflected ray has been reflected $n-1$ times from the surface. The direct ray (*D*) is the zeroth-order bottom-reflected ray.

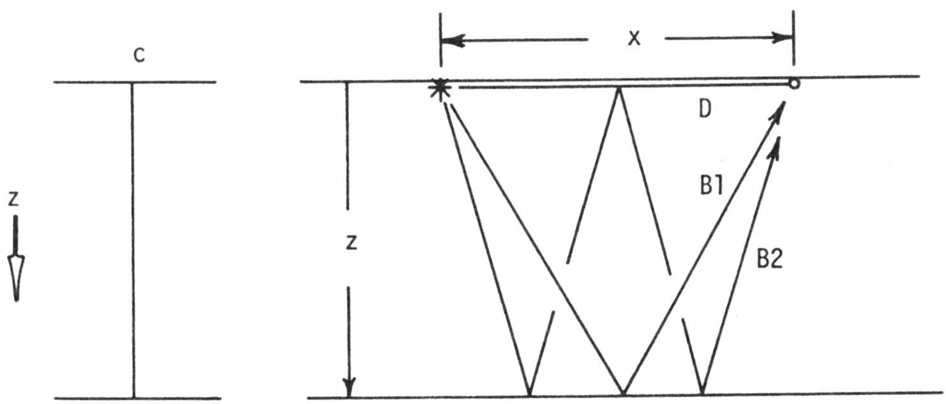

Fig. 51.

The travel time for each ray is given below:

$$t_D = \frac{x}{c} = \frac{[x^2 + (0z)^2]^{1/2}}{c} = \frac{r_0}{c}$$

$$t_{B1} = \frac{[x^2 + (2z)^2]^{1/2}}{c} = \frac{r_1}{c}$$

$$t_{B2} = \frac{[x^2 + (4z)^2]^{1/2}}{c} = \frac{r_2}{c}$$

$$t_{Bn} = \frac{[x^2 + (2nz)^2]^{1/2}}{c} = \frac{r_n}{c}$$

The time sequence of the arrivals of an acoustic pulse is shown in Fig. 52.

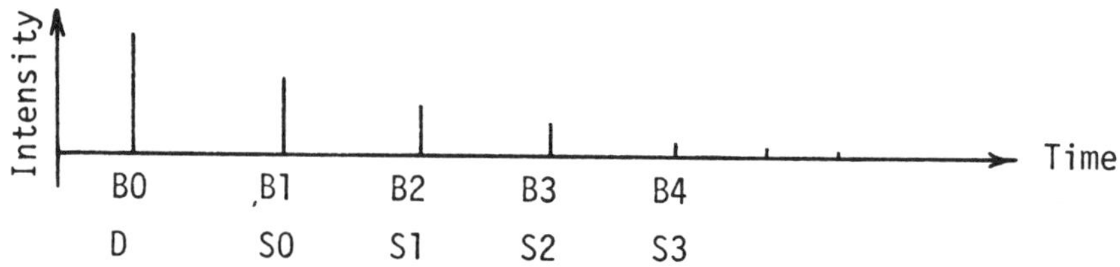

Fig. 52.

Note that there is some indication of a decreasing intensity. Since for each arrival we may assume spherical spreading, intensity will decrease as the inverse square of the propagation distance (r_n). But, in addition to spreading, there will be attenuation and reflection losses. Note that the higher order arrivals will not only have travelled further but will have struck the surfaces more often and that each reflection will have removed more energy than a lower order reflection due to the higher grazing angle.

A plot of the arrival time versus the range x is shown in Fig. 53.

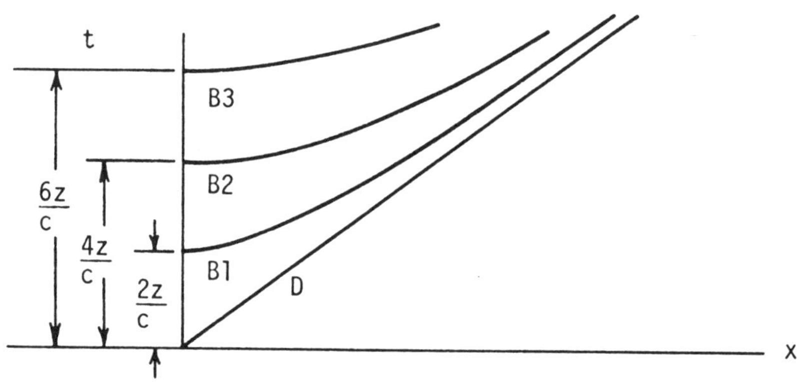

Fig. 53.

These curves are hyperbolas and the t intercepts are the round trip times for vertically reflecting rays.

If we still consider an isovelocity profile, but take the source and receiver at some depth below the surface, we find additional rays present. There will be four rays for each order of bottom-reflected rays: one will be the same as before $[Bn, S(n-1)]$; two distinct rays will make n reflections at the bottom and n reflections at the surface, one strikes the surface first and one strikes the bottom first $2[Bn, Sn]$; and one will reflect from the surface $n+1$ times $[Bn, S(n+1)]$. For the zeroth-order only the two rays $[B0, S0]$ (the direct D) and $[B0, S1]$ (the surface-reflected S) exist. These rays for the zeroth- and first-order bottom-reflected rays are shown in Fig. 54.

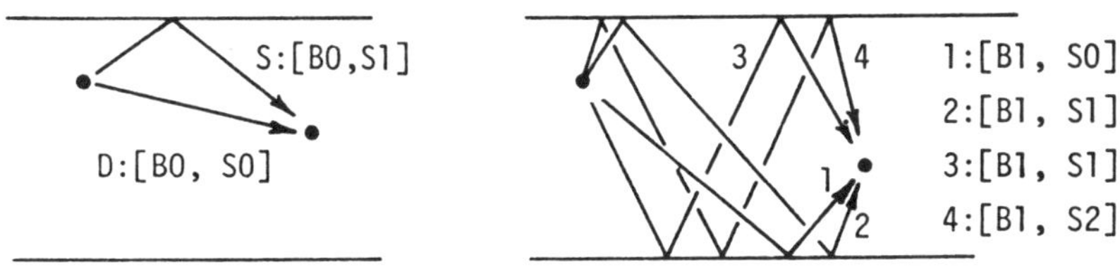

Fig. 54.

The time sequence in Fig. 55 shows their arrivals.

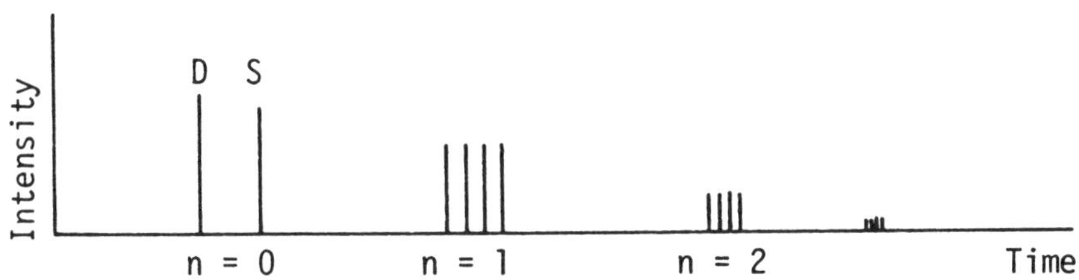

Fig. 55.

For the case of a negative sound-speed gradient, the rays will be refracted toward the bottom as shown in Fig. 56.

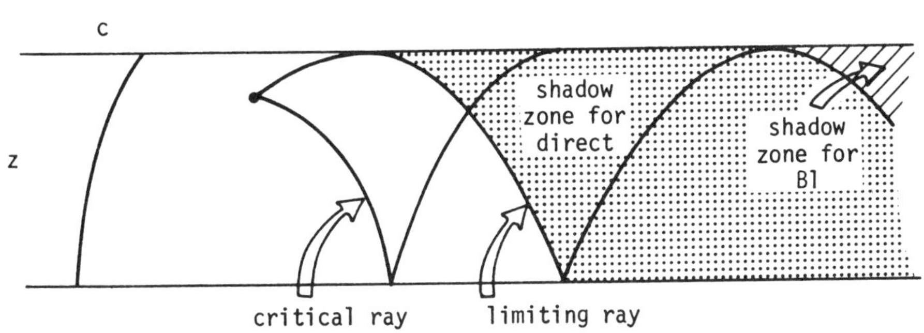

Fig. 56.

In this case no single ray type can propagate indefinitely because each type of ray has a limiting ray beyond which that ray type cannot be found. This establishes the long-range limit for a given ray type. The short-range limit for a given type may be assumed to coincide with the propagation of the critical ray of that ray type. We establish this as the short-range limit since rays incident upon the bottom from angles more normal than the critical ray will not be totally reflected.

Using these rays to delimit propagation zones we obtain the diagram of Fig. 57.

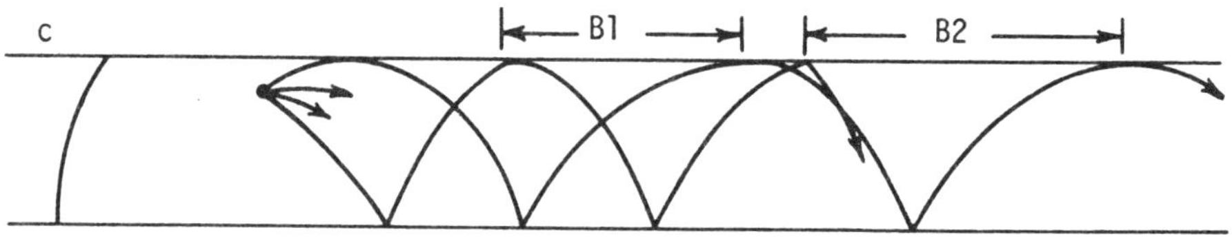

Fig. 57.

Here the interval labeled $B1$ and $B2$ are the zones for propagation of those type rays. The propagation zones may be illustrated on a transmission diagram (see Fig. 58).

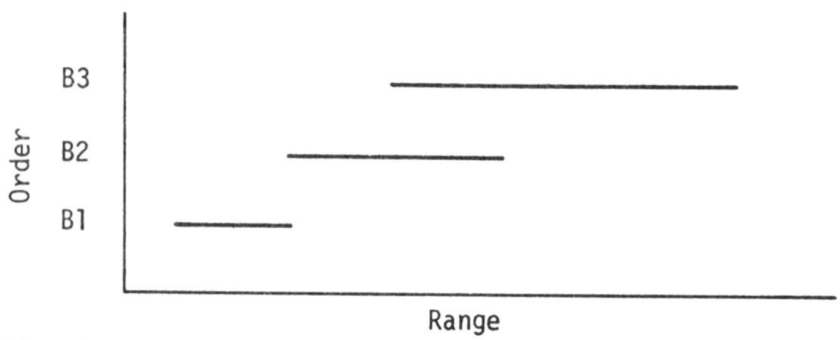

Fig. 58.

In water with parallel boundaries, the width of the range interval for propagation of a Bn ray type is n times the range interval of $B1$.

As shown in Fig. 59, rays are refracted toward the surface in the case of a positive sound-speed gradient.

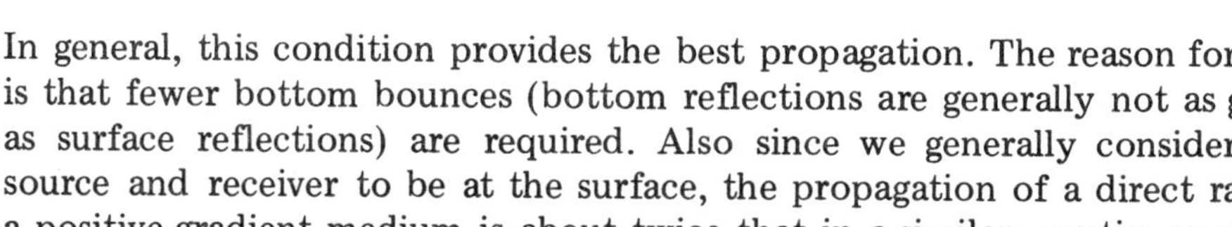

Fig. 59.

In general, this condition provides the best propagation. The reason for this is that fewer bottom bounces (bottom reflections are generally not as good as surface reflections) are required. Also since we generally consider the source and receiver to be at the surface, the propagation of a direct ray in a positive-gradient medium is about twice that in a similar negative-gradient medium. For example, consider Fig. 60.

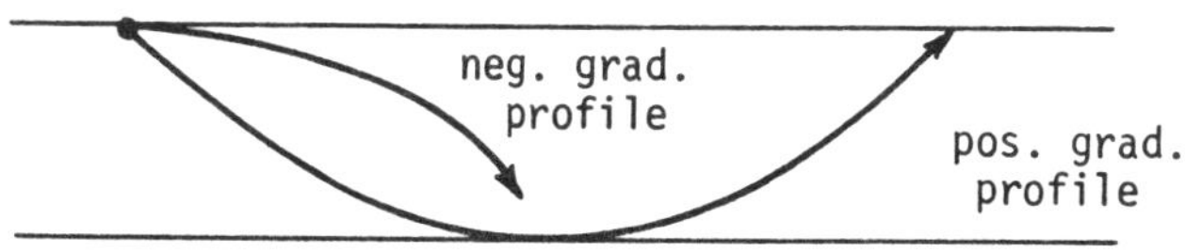

Fig. 60.

For a source and receiver at the bottom the opposite is true. For one at the surface and the other at the bottom considerations such as in Fig. 59 are not important.

In the case of the positive gradient there is a set of rays called "RSR" (refracted-surface-reflected) rays that propagate particularly well. These rays are the rays that are returned to the surface by refraction rather than bottom reflection and therefore do not experience the poor reflective properties of the bottom. The quality of propagation of RSR rays depends upon sea state.

7.3. ATTENUATION OF SOUND IN THE SEA

The attenuation of sound in the sea is characterized by any one or a pair of four attenuation coefficients, α_1, α_2, α_3, or α_4. The frequency ranges of applicability of each of these coefficients are given in Table II.

TABLE II

Coef.	Freq. range	Att. range (dB/kyd)	Process
α_1	500 kHz—up	100—up	viscosity
α_2	10 kHz—500 kHz	1—100	$MgSO_4$ relaxation
α_3	200 Hz— 10 kHz	0.1—1	scattering by inhomog.
α_4	16 Hz—200 Hz	0.001—0.01	possibly a boundary effect

In the center regions of these ranges, a single coefficient is sufficient; near the limits a pair of coefficients should be summed (for example, near 500 kHz $\alpha = \alpha_1 + \alpha_2$).

Useful empirical formulas for these coefficients (in dB/kyd) as a function of frequency (kHz) are given below:

- $$\alpha_1 = 2.68 \cdot 10^{-2} D f^2 / f_T \qquad (7.4)$$

where D is a function of depth and f_T is a function of temperature. These functions are:

$$D = 1 - 1.93 \cdot 10^{-5} d$$

where d is the depth in feet, and:

$$f_T = 2.19 \cdot 10^{7-[1520/(T+273)]}$$

where T is the temperature in °C.

- $$\alpha_2 = 1.86 \cdot 10^{-2} \frac{SDf_T f^2}{f_T^2 + f^2} \qquad (7.5)$$

where D and f_T are given above and S is salinity in parts per thousand.

- $$\alpha_3 = \frac{0.1 f^2}{1 + f^2} \qquad (7.6)$$

- $$\alpha_4 = 0.33 f^2 \qquad (7.7)$$

A plot of the attenuation coefficient over the frequency range 20 Hz to 50,000 kHz is given in Fig. 61.

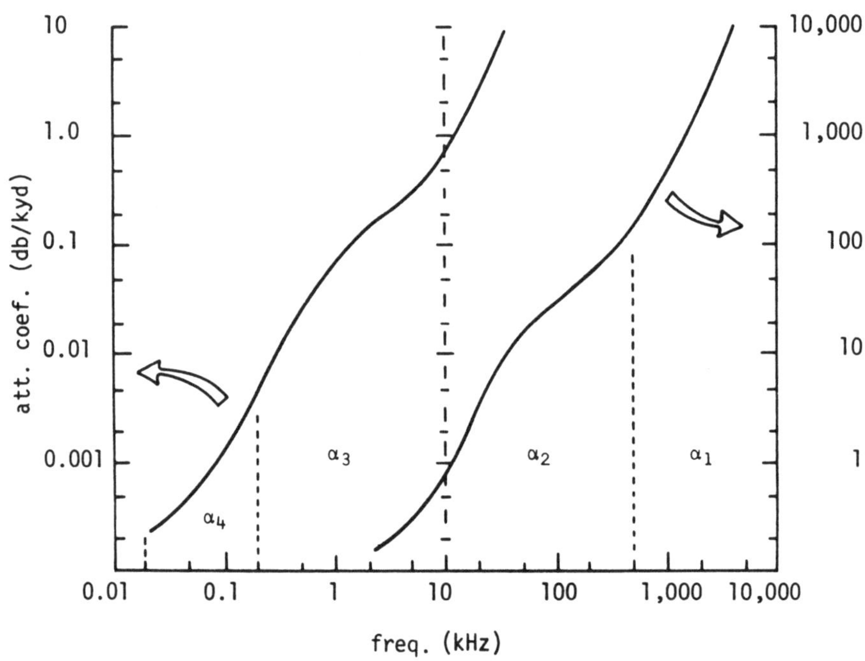

Fig. 61.

CHAPTER 8

REVERBERATION

In addition to the echo return from the target and to ambient noise, an echo-ranging sonar will receive part of its own transmitted energy returned in undesired echos. This undesired return of energy is known as "reverberation". It is the net effect of scattering from various inhomogeneities in the volume of the medium and at the surfaces bounding the medium. Reverberation is proportional to the transmitted power and, therefore, once the return becomes reverberation-limited, instead of noise-limited, no further increase in power or reduction in bandwidth will improve the quality of the return.

Considerable theoretical and analytical formulation has been developed for reverberation phenomena, but it is all rather useless because of complexity and irregularity of the medium. Instead these formulations have provided some rules of thumb for design criteria and order of magnitude prediction models.

The theory is built around two forms of reverberation: (1) volume reverberation, and (2) surface reverberation.

Surface reverberation is further divided into: (a) sea surface reverberation, and (b) bottom reverberation.

These latter two have the same mathematical formulation but differ in physical form.

8.1. THE THEORY OF VOLUME REVERBERATION

In the case of volume reverberation the scattering of sound by inhomogeneities in water may range between two extremes:

(1) Rayleigh scattering, on the one hand, by particles much smaller than a wavelength. This scattering is independent of the shape of the scatterers and depends upon the square of the frequency.

(2) Regular geometric reflection, on the other hand, by objects larger than the wavelength of the sound. This scattering depends upon the acoustical properties of the individual scatterer and is independent of frequency.

The intermediate case of object and wavelength near the same size is more complicated. The function of frequency generally contains resonances that depend upon the acoustical properties of the scatterers.

Some of these inhomogeneities are: marine organisms, entrapped air bubbles, and thermal microstructure.

When the first two do occur, the reverberation they produce is much stronger than that of the last. Reverberation caused by marine organisms is fairly universal and represents the major source of volume reverberation.

To describe volume reverberation in a quantitative manner, we will define several quantities. First, we define "volume scattering coefficient" (m_v) to be the power per unit intensity and scattering volume scattered from an incident plane wave of intensity (I) by a small volume (V), i.e.:

$$m_v = \frac{P_{scat}}{IV} (L^{-1}) \tag{8.1}$$

where P_{scat} is the total power scattered from the beam. Note that m_vV has the dimensions of area and may be interpreted as the effective cross-sectional area of the scattering volume since it intercepts the power $I(m_vV)$. For this reason m_v has also been called the "backscatter cross-section".

We next define the "volume scattering strength" (s_v) to be the intensity per unit incident intensity and scattering volume scattered from an incident plane wave by a small volume and measured at a unit reference distance in the direction (θ, ϕ), i.e.:

$$s_v(\theta, \phi) = \frac{I_{scat}}{IV} (\theta, \phi) (L^{-3}) \tag{8.2}$$

where $I_{scat}(\theta, \phi)$ is the intensity scattered from the beam into the direction (θ, ϕ) and measured at the reference distance from the acoustic center of the scatterer.

"Volume backscattering strength ($s_v^{(b)}$) is the scattering strength for the backscatter direction.

Since the power may be obtained by integrating the intensity over the surface of a sphere, i.e.:

$$P_{scat} = \int I_{scat} r_0^2 \, d\Omega$$

where r_0 is the reference distance and is unity:

$$m_v = \int s_v \, d\Omega \tag{8.3}$$

It is generally assumed that:

$$s_v = s_v^{(b)}$$

for all angles (i.e., the scattering is isotropic).*

* Probably valid only in the case of resonant scattering.

In this case:

$$m_v = 4\pi s_v^{(b)} = 4\pi s_v \tag{8.4}$$

The "decibel volume scattering strength" is given by:

$$S_v = 10 \log s_v \tag{8.5}$$

By the assumption of isotropy above:

$$S_v = 10 \log \frac{m_v}{4\pi} \tag{8.6}$$

The total scattering strength of an insonified volume (V) of ocean is:

$$\frac{I_{scat}}{I} = \int_V s_v \mathrm{d}V$$

Usually s_v is assumed constant in the volume so that the integral reduces to $s_v V$.

The decibel value of the total scattering strength is equivalent to the target strength so that:

$$TS = 10 \log \frac{I_{scat}}{I} = S_v + 10 \log V \tag{8.7}$$

The "reverberation level" for volume reverberation (RL_v) is the level of a plane wave, incident along the acoustic axis, that produces the same hydrophone response as the reverberation. The reverberation level, for the case of the projector and hydrophone being the same transducer, is:

$$RL_v = SL - 2TL + S_v^{(b)} + 10 \log V \tag{8.8}$$

where SL is the source level, $2TL$ is the two-way transmission loss, $S_v^{(b)}$ is the back scattering strength and V is the effective insonified volume.

The effective volume is given by:

$$V = r^2 \left(\frac{c\tau}{2}\right) \int_0^{4\pi} b(\theta, \phi)\, b'(\theta, \phi)\, \mathrm{d}\Omega \tag{8.9}$$

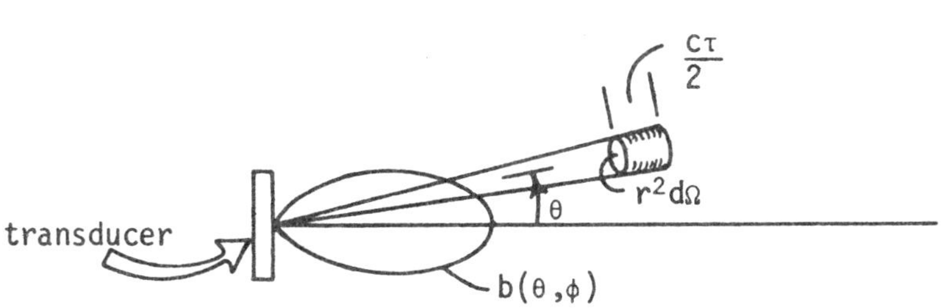

Fig. 62.

where τ is the pulse length and b and b' are the transmit and receive beam patterns, respectively. This is easily understood by referring to Fig. 62. Let us define an equivalent solid angle beamwidth Ψ such that:

$$\Psi = \int_0^{4\pi} b(\theta, \phi)\, b'(\theta, \phi)\, d\Omega \tag{8.10}$$

Assuming spherical spreading, equation 8.8 may be written:

$$RL_v = SL - 40 \log r + S_v^{(b)} + 10 \log \frac{r^2 c\tau}{2} + 10 \log \Psi \tag{8.11}$$

Expressions for $10 \log \Psi$ (as well as $10 \log \Phi_0$ to be defined shortly) are given in Table III.

TABLE III

Array	$10 \log \Psi$ (dB re 1 steradian)	$10 \log \Phi_0$ (dB re 1 radian)
Circular plane array, in an infinite baffle of radius $a > 2\lambda$	$20 \log \frac{\lambda}{2\pi a} + 7.7$	$10 \log \frac{\lambda}{2\pi a} + 6.9$
Rectangular array in an infinite baffle, side a horizontal, b vertical, with $a, b \gg \lambda$	$10 \log \frac{\lambda^2}{4\pi ab} + 7.4$	$10 \log \frac{\lambda}{2\pi a} + 9.2$
Horizontal line of length $1 > \lambda$	$10 \log \frac{\lambda}{2\pi l} + 9.2$	$10 \log \frac{\lambda}{2\pi l} + 9.2$
Nondirectional (point) transducer	$10 \log 4\pi = 11.0$	$10 \log 2\pi = 8.0$

From P.A. Barakos, *Underwater Reverberation as a Factor in ASW Acoustics*, U.S. Navy Underwater Sound Lab. Rept. 620, Sept., 1964.

8.2. THE THEORY OF SURFACE REVERBERATION

By surface reverberation we mean the reverberation produced by the scattering of acoustical energy from surfaces rather than in the volume of the medium. Naturally, the two surfaces that we are concerned with in underwater acoustics are the sea surface and the sea floor.

We will define surface reverberation quantities in a manner similar to that which was done in volume reverberation. The quantities are:

— "Surface scattering coefficient" m_s — the power per unit intensity and scattering surface area scattered from an incident plane wave of intensity (I) by a small surface (A), i.e.:

$$m_s = \frac{P_{scat}}{IA} \text{ (dimensionless)} \tag{8.12}$$

— "Surface scattering strength" s_s — the intensity per unit incident intensity and scattering surface area scattered from an incident plane wave by a small area and measured at a unit reference distance in the direction (θ, ϕ), i.e.:

$$s_s = (\theta, \phi) = \frac{I_{scat}(\theta, \phi)}{IA} \; (L^{-2}) \tag{8.13}$$

— "Decibel surface scattering strength" S_s — defined by:

$$S_s = 10 \log s_s \tag{8.14}$$

Integrating the scattered intensity over the surface of a unit sphere gives the scattered power as before. The equation can then be reduced to:

$$m_s = \int_0^{2\pi} s_s \mathrm{d}\Omega \tag{8.15}$$

Making the assumption:

$$s_s(\theta, \phi) = s_s^{(b)}$$

for all (θ, ϕ) in the upper half-space*, we obtain:

$$m_s = 2\pi s_s^{(b)} = 2\pi s_s \tag{8.16}$$

The reverberation level is:

$$RL_s = SL - 2TL + S_s^{(b)} + 10 \log A \tag{8.17}$$

where A is the effective reverberation area. The expression for the effective reverberation area is considerably more difficult to interpret than for the reverberation volume (equation 8.9). The reason for this is that geometry of the scattering surface in relation to the position and orientation of the transducer must be considered.

Let the angles α and β represent the angles to a point (P) on the surface relative to the position of the transducer (T).

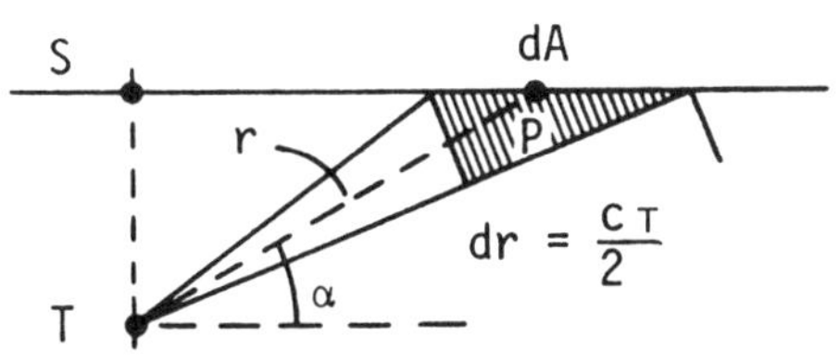

Fig. 63. β is the angle around the line TS.

* It is most unlikely that this assumption is ever valid in any strict sense.

From the diagram (Fig. 63) we find that the elemental area ($\mathrm{d}A$) is given by:

$$\mathrm{d}A = \mathrm{d}x(x\mathrm{d}\beta) = x\mathrm{d}x\mathrm{d}\beta$$

We also see that:

$$x\mathrm{d}x = r\mathrm{d}r = r\frac{c\tau}{2}$$

Let $\beta = \phi$, i.e., chose β so that it coincides with the azimuthal angle about the transducer, then:

$$\mathrm{d}A(\alpha) = r(\alpha)\frac{c\tau}{2}\mathrm{d}\phi$$

Define the effective elemental area $\mathrm{d}A'$ to be the elemental area weighted by the composite beam pattern:

$$\mathrm{d}A'(\alpha, \theta, \phi) = r(\alpha)\frac{c\tau}{2}b(\theta, \phi)\, b'(\theta, \phi)\, \mathrm{d}\phi$$

where θ now brings in the vertical orientation of the transducer. See also Fig. 64.

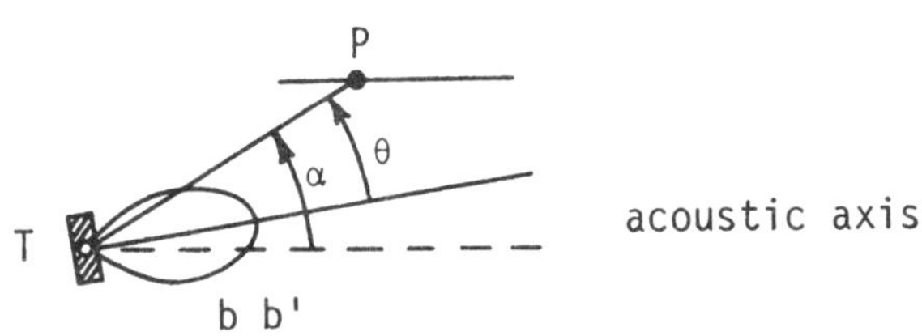

Fig. 64.

The effective insonified area at the point P is:

$$A(\alpha, \theta) = r(\alpha)\frac{c\tau}{2}\int_0^{2\pi} b(\theta, \phi)\, b'(\theta, \phi)\, \mathrm{d}\phi \qquad (8.18)$$

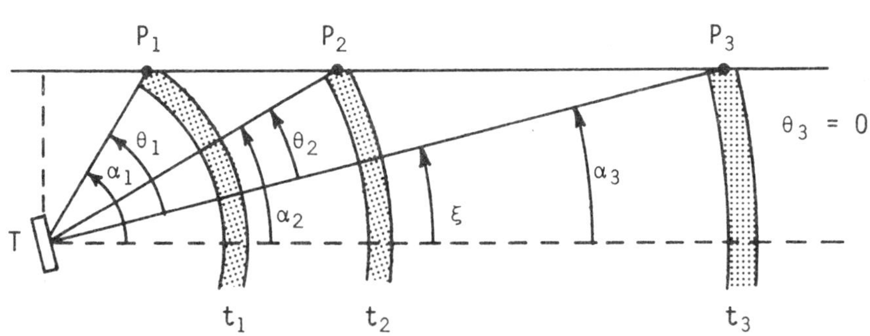

Fig. 65.

Note that, unlike the effective volume given in (8.9), the effective area has angular dependences. It should be remembered that the range r (or the angle α) is not connected geometrically to the angle θ until the orientation of the transducer is specified. See the details in Fig. 65. Let:

$$\Phi(\theta) = \int_0^{2\pi} b(\theta, \phi)\, b'(\theta, \phi)\, d\phi \tag{8.19}$$

then:

$$A(\alpha, \theta) = r(\alpha) \frac{c\tau}{2} \Phi(\theta) \tag{8.20}$$

This is then put into (8.17) to obtain:

$$RL_s = SL - 40 \log r + S_s^{(b)} + 10 \log \left(\frac{rc\tau}{2}\right) + 10 \log \Phi(\theta) \tag{8.21}$$

It may be shown that $\Phi(\theta)$ can be written:

$$\Phi(\theta) = \left[\frac{b(\theta - \xi, 0)\, b'(\theta - \xi, 0)}{\cos \theta}\right] \Phi_0 \tag{8.22}$$

where ξ is the elevation angle of the transducer axis (see Fig. 65) and θ is less than 30°, and Φ_0 is given by:

$$\Phi_0 = \Phi(0) = \int_0^{2\pi} b(0, \phi)\, b'(0, \phi)\, d\phi$$

and is tabulated at the end of the last section. We may assume that outside 30° the beam patterns are down so low that reverberation may be neglected. The correction factor (F) for $\xi = 0$ is:

$$F = 10 \log \frac{b(\theta, 0)\, b'(\theta, 0)}{\cos \theta}$$

For a circular or rectangular piston it is given approximately by Table IV.

TABLE IV

θ (degrees)	F (dB)
0	0
2	−1
4	−2
6	−6
8	−12

The reverberation level (equation 8.21) becomes:

$$RL_s = SL - 40 \log r + S_s + 10 \log \left(\frac{c\tau}{2} r\right) + 10 \log \Phi_0 + F \tag{8.23}$$

if the acoustic axis of the transducer is horizontal.

Lambert's law

A particular type angular scattering distribution which has found some use in optics and acoustics satisfies a certain rule known as Lambert's law. Accordingly, the scattering distribution obeys:

$$I_{scat} = \mu I A \sin \alpha \sin \beta \tag{8.24}$$

where I_{scat} is the scattered intensity as a function of the grazing incident angle (α) and scattered angle (β), I is the incident intensity, A is the reverberation area, and μ is a proportionality constant (which is $1/\pi$ if no energy is lost into the medium below).

This distribution is plotted in Fig. 66.

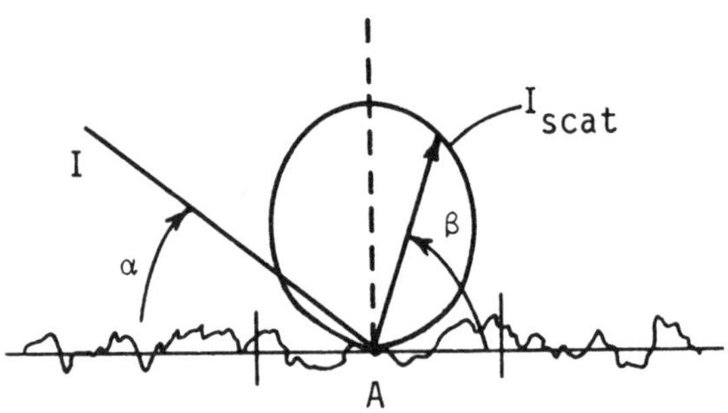

Fig. 66.

The scattering strength as a function of α and β is:

$$S_s(\alpha, \beta) = 10 \log \frac{I_{scat}}{IA} = 10 \log \mu + 10 \log (\sin \alpha \sin \beta)$$

The decibel back scattering strength is:

$$S_s^{(b)} = 10 \log \mu + 10 \log (\sin^2 \alpha) \tag{8.25}$$

At normal incidence:

$$S_s^{(b)} = 10 \log \mu$$

and for total reflection:

$$S_s^{(b)} = -10 \log \pi = -5 \text{ dB}$$

This is a very special form of scattering but very rough surfaces seem to follow it fairly well.

8.3. REVERBERATION AS OBSERVED AT SEA

Volume reverberation

The major source of volume reverberation in sea is the deep scattering layer(s), DSL as it is called. The scatterers responsible for the reverberation are undoubtedly biological, but the specific creatures making up the DSL have not been definitely identified.

Many studies using towed nets, photography, and manned submersibles as well as echo-ranging have provided only one definite result; the number of creatures per cubic meter responsible for even a strong DSL is quite small ($\simeq 0.05\,\mathrm{m}^{-3}$). The studies suggest that the organisms most likely involved are myctophids (lantern fish), siphonophores, euphausids (shrimp like creatures), squid, and copepods.

We may summarize the acoustic characteristics of the DSL as follows:

— The DSL is fairly universal.

— At any one time and location there may be several layers at different depths.

— Some layers, but not all, migrate upward at sunset and downward at sunrise.

— Their depths are more or less constant during the daytime, and range from a few tens of fathoms to as many as a thousand.

— A variable and unpredictable frequency structure is found between 1 and 20 kHz generally indicating resonance phenomena.

— Above 20 kHz there is a general increase in scattering strength of 3—5 dB/octave.

— Some layering has been observed in shallow water.

Sea-surface reverberation

Sea-surface reverberation is one of the more important effects on sonar operation because of the shallow depth of the transducer in the surface ship. This reverberation is extremely important in surface channel propagation.

It is known that the sea-surface back scattering strength varies with incident angle, acoustic frequency, and roughness of the surface. The latter is often related to the wind speed in a somewhat semi-quantitative way.

In addition to the actual reflection and scattering that take place at the air—sea interface, a volume scattering due to near-surface bubbles occurs within a foot of the surface. Since this is so close to the surface and can not be distinguished from actual surface scattering except by a detailed interpretation of the data, it is considered as part of the surface reverberation.

Several empirical and semi-theoretical expressions have been advanced to provide the functional dependence of back scattering strength upon incident angle, acoustic frequency, and wind speed. None succeed completely.

One such expression was given by Eckart* and reduced to a practical form by Chapman and Scott**. It is:

$$S_s^{(b)} = -10 \log 8 \pi \alpha^2 + 2.17 \alpha^{-2} \tan^2 \theta$$

where θ is the grazing angle and α^2 is the mean-square slope of the surface waves. α^2 is given by Cox and Munk*** to be:

$$\alpha^2 = 0.003 + 5.12 \cdot 10^{-3} W$$

where W is the wind speed in meters per second. Good agreement was found for angles greater than 60° and no frequency dependence was observed as Eckart's theory indicates for large grazing angles. But this formula is not valid very near normal incidence.

It would seem intuitively obvious that for low grazing the backscatter should decrease to extremely low values independent of the sea state, but experimentally it does not. It is strongly dependent upon sea state and frequency. It is to account for this behavior that the very near-surface scattering layer of bubbles was conjectured.

A sketch of the grazing angle dependences for a low sea state (L) and a high sea state (H) is shown in Fig. 67.

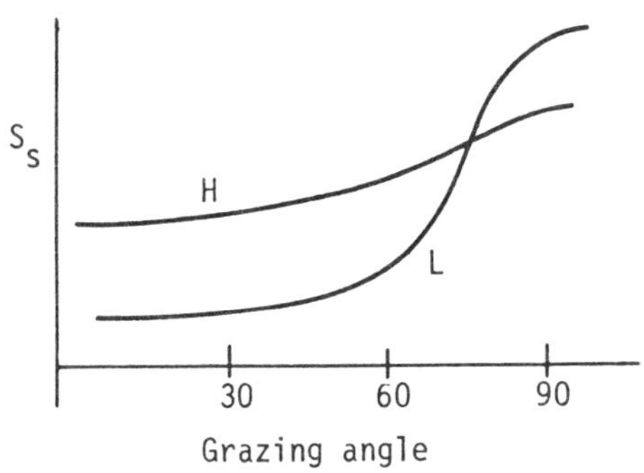

Fig. 67.

Bottom reverberation

In shallow water, scattering from the bottom is generally the strongest contributor to the observed reverberation. Typical order of magnitude values for volume, sea surface, and bottom reverberation in shallow water are − 80 dB, − 40 dB, and − 25 dB, respectively.

---

* C.E. Eckart, *Scattering of Sound from the Sea Surface*. J. Acoust. Soc. Am., **25**: 566 (1953).

** R.P. Chapman and H.D. Scott, *Surface Backscattering Strengths Measured over an Extended Range of Frequencies and Grazing Angles*. J. Acoust. Soc. Am., **36**: 1735 (1964).

*** C. Cox and W. Munk, *Measurements of Roughness of the Sea Surface from Photographs of the Sun's Glitter*. J. Opt. Soc. Am., **44**: 838 (1954).

Bottom reverberation depends upon the type and coarseness of the bottom and its contours. But it is generally believed that the contours produce the biggest effect. This is confirmed by an absence of a strong frequency dependence for frequencies below 10 kHz. For scattering at frequencies above 10 kHz, the bottom properties may become more significant. Finer bottoms such as those composed of sand, silt, and mud show an increase in scattering strength of 3 dB per octave increase in frequency, while rock and sand mixed with rock and shell bottoms seem to show no frequency dependence up to at least 60 kHz.

CHAPTER 9

NOISE

Generally, noise observed in the sea is divided into three categories: that generated by the measuring process itself, e.g., cable strumming, platform noise, etc., this noise is called "self-noise"; anomalous localized noise due to individual fish, ships, etc.; and ambient noise, the general background noise that is left when the specific sources indicated above are removed.

The sonar parameter "noise level" (NL) is the intensity level as measured by a hydrophone operating in a specific bandwidth. Noise level is expressed in decibels with a reference of a plane wave of 1 dyn/cm^2 rms pressure. The level cited in literature is generally a spectrum level (per unit bandwidth) in order that the effect of the bandwidth may be eliminated. The noise level is then given by:

$$NL = NL_{1\,\mathrm{Hz}} + 10 \log W \qquad (9.1)$$

where W is the bandwidth.

There are several major sources of ambient noise and each dominates portions of the frequency spectrum. Spectra have been measured over the range from a few Hz to a few hundred kHz. In this range the major sources of ambient noise are:

radiated sound associated with waves, surf, rain, etc.;
pressure changes due to oceanic turbulence;
distant shipping and industrial activity;
marine animals as a group rather than localized individuals; and thermal molecular activity.

Other phenomena that generate acoustic noise in the sea are the hydrostatic pressure changes due to tides and waves and seismic activity, but generally this noise is below one Hz and not significant in underwater acoustic applications.

TABLE V

Frequency (Hz)	Process
1–10	Oceanic turbulence
10–200	Shipping
200–50,000	Wind, waves, foam, and spray
50,000–100,000	Molecular activity

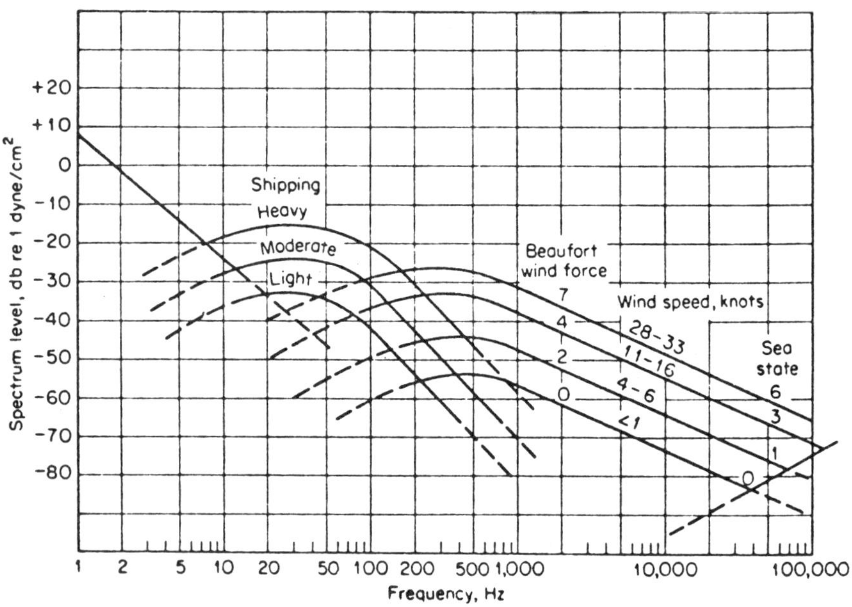

Fig. 68.

Average deep-water ambient noise spectra can be divided into four distinct regions as shown in Fig. 68.*

Table V indicates the frequency ranges and the dominant process.

Due to increased biological, human, and natural activity, shallow-water ambient noise is much more variable than deep-water ambient noise but, in general, the average shallow-water ambient noise is about 9 dB higher than deep-water ambient noise.

The directivity pattern of sea surface noise is not definitely known, but it is generally assumed to be of the form $\cos^m \theta$, where $m = 0$, 1, or 2. These angular dependences are plotted in Fig. 69.

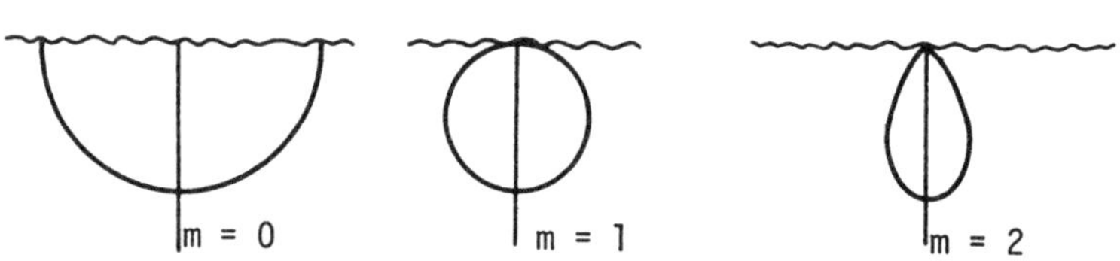

Fig. 69.

* From *Principles of Underwater Sound for Engineers*, by R.J. Urick. 1967, McGraw-Hill, Inc. Used with permission of McGraw-Hill Book Company.

It has been postulated that in the above cases a decrease in ambient noise with depth is expected due primarily to attenuation. Accordingly, it is expected that there would be a greater depth dependence for high frequency than for low. Experiments generally confirm this.

Noise in the ocean is found to be nonisotropic with the low frequencies coming predominantly from the horizontal and the high frequencies from the surface as shown in Fig. 70.

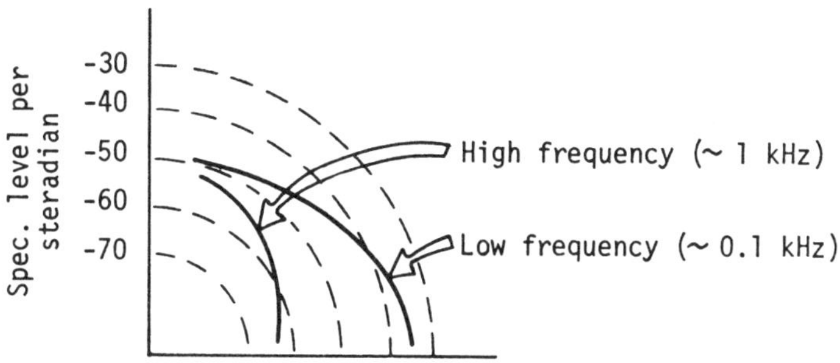

Fig. 70.

This is in agreement with both the phenomenon of high-frequency attenuation and the dominance of the low-frequency spectrum by distant shipping.

CHAPTER 10

SONAR SIGNAL PROCESSING

When a signal is weak, the information it contains can be interpreted only after it is enhanced relative to background noise by a signal processor. The theory upon which optimum signal processors are based is complex. Only the fundamentals of the theory are discussed in this chapter. Hardware implementation and mathematical details are beyond the scope of this book.

10.1. DETECTION THEORY

Detecting a signal in the presence of background noise is essentially a decision process. For any signal input to the receiver, a target signal is either present (hypothesis H_1) or absent (hypothesis H_0). For this input a decision must be made that a signal is in fact present (detection) or that it is absent (dismissal). The four possible outcomes of the two hypotheses and the two decisions are shown below.

Hypotheses	Decision	
	Detection	Dismissal
Signal absent (H_0)	False detection	Correct dismissal
Signal present (H_1)	Correct detection	False dismissal

Obviously, correct decisions are desired, but they are not always possible. This is especially true when the signal is weak or when it is not present at all, or when the period of time in which to make the decision is limited. False detections are costly in terms of effort wasted in pursuing the false alarms and false dismissals are costly in terms of the presence of a rejected true target.

The purpose of the signal processor is to improve the prospects for making correct decisions. The optimum signal processor is an ideal device which operates on the input signal in such a way as to yield the theoretically best performance for a given input signal-to-noise ratio.

The action of the basic optimum signal processor is threefold: (1) it operates on the input signal to form an estimate of the optimum test statistic, $\hat{\alpha}$;* (2) it forms the likelihood ratio (or a related parameter) for the estimated

* For a listing of symbols used in this chapter see the Notation.

NOTATION Detection theory quantities and formulas

Quantity	Symbol	Formula
Hypothesis that a signal is present	H_1	
Hypothesis that a signal is not present	H_0	
Target signal	$s(t)$	
Ambient noise	$n(t)$	
Signal input to processor	$x(t)$	$= s(t) + n(t)$
Test statistic	α	
Estimate of test statistic	$\hat{\alpha}$	
Threshold of test statistic	α_0	
Probability density function at α given $H_{i=0,1}$	$p(\alpha : H_i)$	
Likelihood ratio	$l(\alpha)$	$= p(\alpha : H_1)/p(\alpha : H_0)$
Estimate of likelihood ratio	$\hat{l}$	$= l(\hat{\alpha})$
Threshold of likelihood ratio	l_0	$= l(\alpha_0)$
Probability of detection given H_1	P_D	$= \int_{\alpha_0}^{\infty} p(\alpha : H_1)\, d\alpha$
Probability of false alarm given H_0	P_{FA}	$= \int_{\alpha_0}^{\infty} p(\alpha : H_0)\, d\alpha$
Mean value of α given H_1	μ_{sn}	
Mean value of α given H_0	μ_n	
Mean difference	m	$= \mu_{sn} - \mu_n$
Standard deviation	σ	
Detection index	d	$= (m/\sigma)^2$
Total energy in the time period 0 to T	E	$= \int_0^T x^2(t)\, dt$
Noise in a 1 Hz band	N_0	
Signal-to-noise ratio	S/N	
Correlation function	$R(t)$	$= \frac{1}{T}\int_0^T x(t)\, s(t - t')\, dt'$

test statistic, $l(\hat{\alpha}) = \hat{l}$; and (3) it compares the calculated likelihood ratio to the preset threshold likelihood ratio, l_0. The decision maker, man or machine, accepts hypothesis H_1 if the calculated value is equal to or exceeds the preset threshold value; it accepts hypothesis H_0 if the calculated value is less. Figure 71 shows a diagram of the optimum processor. The two graphs

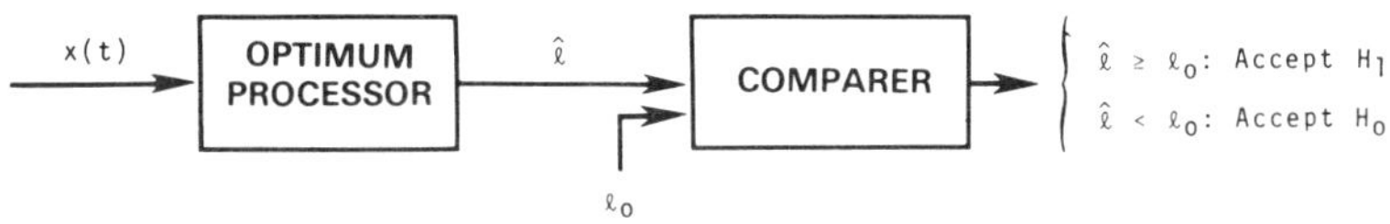

Fig. 71.

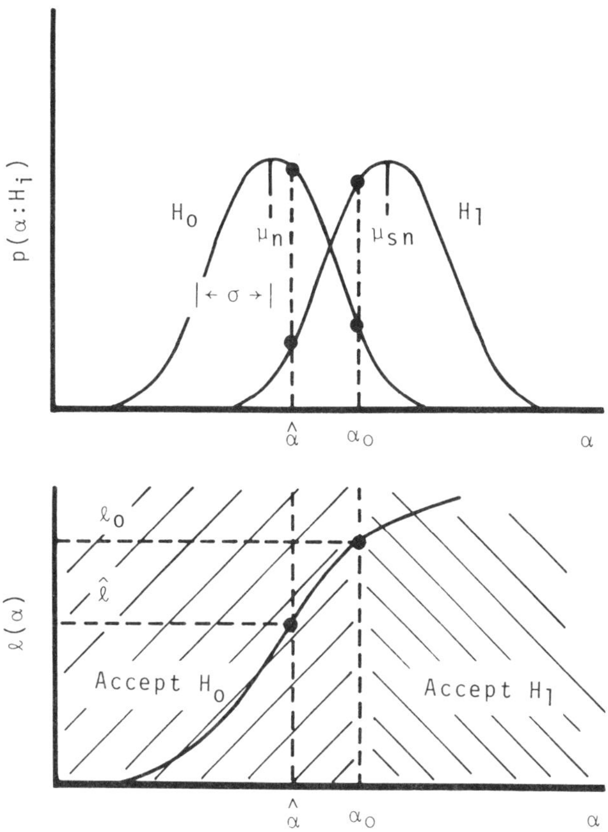

Fig. 72.

in Fig. 72 show the relationship among the test statistic, the probability distribution for each of the two hypotheses, and the likelihood ratio. Also shown in the lower graph are the domains for acceptance of H_0 and H_1.

The choice of the best test statistic is determined by how much is known about the signal being sought and about the background noise. Generally, for passive detection little is known about the signal, except possibly the band of frequencies, more or less narrow, in which the signal is concentrated. In the case of coherent, narrowband lines radiated from machinery, a signal known except for phase over a more or less long time interval may be assumed. Usually more is known about active sonar signal echos because they are to some extent replicas of the known transmitted signals. For simplicity a Gaussian distribution and white (flat) spectrum are assumed often for the background noise.

To clarify these points three examples of optimum processors are discussed below.

Example A (Energy Processor): The target signal is unknown except for the band of frequencies in which it is concentrated (f to $f + \Delta f$). The ambient noise is assumed to be white and Gaussian. The optimum processor forms an estimate of the energy in the signal band for the time period 0 to τ, i.e.:

$$\hat{E} = \int_0^\tau x'(t)^2 \, dt \tag{10.1}$$

where $x'(t)$ is the input signal, $x(t)$, filtered in the target signal band. This processor is illustrated in Fig. 73.

Fig. 73.

Example B (Quadrature Processor): The target signal is known to be stable in phase and not strongly varying in amplitude for the period τ and to have an angular frequency ω, but the value of the phase is unknown. The signal is given by:

$$s(t) = a(t) \cos(\omega t + \theta)$$

Figure 74 shows the optimum processor for this case.

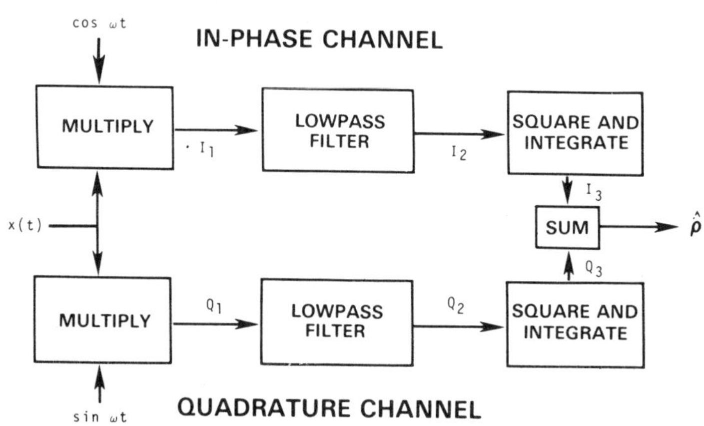

Fig. 74.

The output of the quadrature processor is approximated by:

$\hat{\rho} = \overline{a^2(t)}$ + small noise terms

The development of this result and the specification of the signals at various points in the diagram are left for a student problem at the end of this chapter.

Example C (Correlation Processor): The target signal, $s(t)$, is completely known and the background noise is uncorrelated. In this case the optimum processor forms an estimate of the correlation function, i.e.:

$$\hat{R}(\tau) = \frac{1}{T}\int_0^T x(t)\, s(t-\tau)\,\mathrm{d}t \tag{10.2}$$

Figure 75 shows the optimum processor for this case.

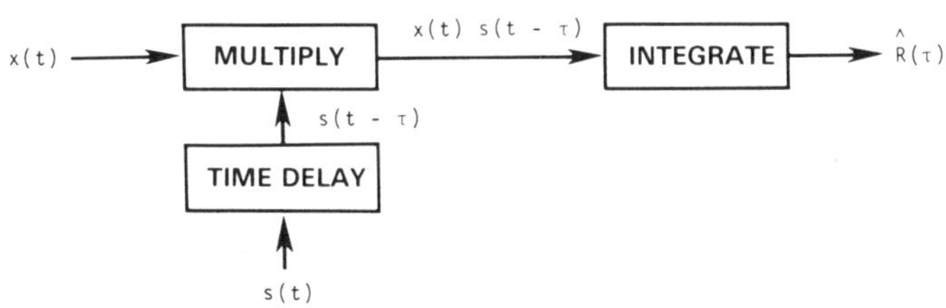

Fig. 75.

Example A is a case of incoherent processing, while examples B and C are cases of coherent processing. Coherent and incoherent processing are further discussed later.

10.2. PROBABILITIES OF DETECTION AND FALSE ALARM

The confidence that can be placed in the validity of a decision is an important measure of the performance of a processor. This confidence can be measured in terms of the probability of detection given hypothesis H_1 (P_D) and the probability of false alarm given hypothesis H_0 (P_{FA}). The probabilities, P_D and P_{FA}, are the areas under the probability distribution curves $p(\alpha:H_1)$ and $p(\alpha:H_0)$, respectively, taken from a value of α to infinity. The value of α of greatest interest is α_0 because it is the critical point in the decision process. Figure 76 shows the above points graphically.

In general, calculating these probabilities is extremely difficult. However, if the distributions are established on the basis of a large number (greater than 20) of independent samples of the test statistic, the distribution becomes Gaussian. Moreover, if the signal is nonfluctuation, the two distributions are identical except that their statistical means are displaced by a distance m.

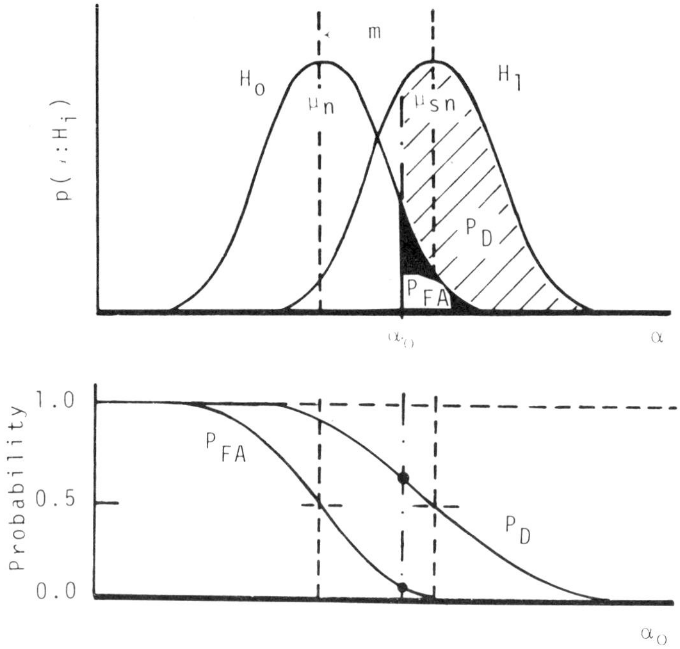

Fig. 76.

It is not difficult to show that, in this case, these probabilities reduce to:

$$P_D = \tfrac{1}{2}\,\text{erfc}\,[(t_0 - \sqrt{d})/\sqrt{2}] \quad (10.3)$$

and:

$$P_{FA} = \tfrac{1}{2}\,\text{erfc}\,[t_0/\sqrt{2}] \quad (10.4)$$

where erfc is the complementary error function which may be found tabulated in standard mathematics handbooks, and:

$$t_0 = (\alpha_0 - \mu_n)/\sigma$$

Note that t_0 is a measure of detection threshold. Deriving equations 10.3 and 10.4 are left for student problems.

In the defining equation for the detection index, d, it is clear that the numerator is related to the signal energy and the denominator is related to the noise energy. Therefore, d is a measure of the input signal-to-noise ratio, S/N. To avoid an involved mathematical development, the proper relationships between d and S/N are simply stated below. For coherent processing:

$$d = 2\Delta f \tau \frac{S}{N} \quad (10.5)$$

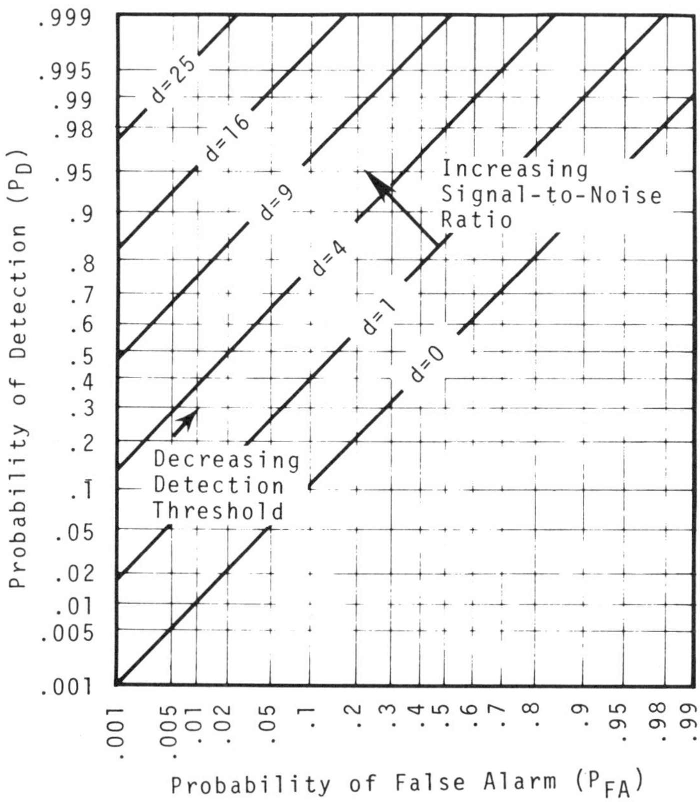

Fig. 77.

For incoherent processing:

$$d = \Delta f\tau \left(\frac{S}{N}\right)^2 = \frac{\tau}{\Delta f}\left(\frac{S}{N_0}\right)^2 \tag{10.6}$$

For a given value of d, P_D and P_{FA} are parametrically represented in the parameter t_0. This representation is shown in Fig. 77. The specification of the desired probabilities of detection and false alarm for a given value of d establishes the minimum detectable input signal-to-noise ratio (detection threshold in the next section). Curves such as those shown in Fig. 77 were first developed by Peterson and Birdsall in 1953 and are called receiver—operator characteristic (ROC) curves.*

---

* Peterson, W.W. and Birdsall, T.G., 1953. *The Theory of Signal Detectability.* Engineering Research Institute, University of Michigan, Technical Report 13.

10.3. DETECTION THRESHOLD

In evaluating the detection performance of a sonar, the signal processor's contribution to the sonar equation (see Chapter 5) is brought in through a system parameter known as the "detection threshold". Detection threshold is defined as the decibel level of the minimum detectable input signal-to-noise ratio (signal level is in the analysis band, Δf, and noise level is in a unit hertz bandwidth). The defining equation is:

$$DT = 10 \log (S/N_0) \tag{10.7}$$

Substituting the detection index for the signal-to-noise ratio using equations 10.5 and 10.6, we find for coherent processing:

$$DT_{\mathrm{coh}} = 10 \log (d/2\tau) \tag{10.8}$$

and for incoherent processing:

$$DT_{\mathrm{inc}} = 5 \log (d\, \Delta f/\tau) \tag{10.9}$$

Since the smaller the minimum detectable level the better the processor performance, we expect the better processors to have small or even negative values of the detection index.

10.4. NARROWBAND PROCESSING

Narrowband filtering has long been the basic operation in signal analysis. It has been used in conventional analyzers to improve signal-to-noise ratio, e.g., in energy detection and to display a frequency spectrum to allow for signal classification. In advanced, modern processors executing various algorithms* for additional processing gain, high-resolution frequency filtering, i.e., Fourier transforming, is still the basic operation. The reason the processor first executes a Fourier transform is because it, like the mathematics it follows, operates more easily in the frequency domain than in the time domain. Today even correlation analysis is accomplished faster in the frequency domain.

The purpose of the narrowband filter is to provide an estimate of the amplitude of the signal component in the frequency band from f to $f + \Delta f$, while rejecting all other bands. For each time period τ_0, where:

$$\tau_0 = 1/\Delta f \tag{10.10}$$

an independent estimate of the energy may be obtained. If each estimate is actually obtained on the basis of a longer period, τ, or by accumulating $\nu/2$ independent estimates, where:

$$\tau = (\nu/2)\tau_0 = \nu/2\Delta f \tag{10.11}$$

---

* Machine operations following the steps of a mathematical formulation.

greater confidence may be placed in it. A set of such estimates is distributed according to the chi-square distribution having ν degrees of freedom. For $\nu > 20$ the distribution is approximately Gaussian. The number of degrees of freedom is twice the time—bandwidth product, i.e., $\nu = 2\Delta f \tau$.

The confidence in an estimate can be calculated from chi-square statistics. Figure 78 shows a common plot of the confidence versus the number of degrees of freedom. Note, for example, that for twenty degrees of freedom (time—bandwidth product of 10) 90% confidence may be placed in the statement that the estimated value is between 0.55 and 1.6 times the true value. As the number of degrees of freedom increases the confidence in the estimate becomes better.

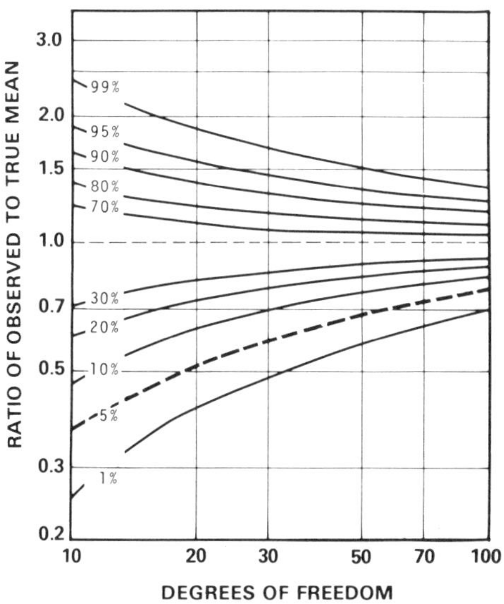

Fig. 78.

PROBLEMS

1. Derive an expression for each of the signals I_1, I_2, I_3, Q_1, Q_2, and Q_3 shown in the diagram of the quadrature processor. Show that the output of the processor is approximately given by:

$$\hat{\rho} = \overline{a^2(t)} + \text{small noise terms.}$$

2. Derive equations 10.3 and 10.4.

GENERAL REFERENCE BOOKS

Albers, V.M. (editor), 1963. *Underwater Acoustics: Proceedings of N.A.T.O. Sponsored Institute at Imperial College, July 31—Aug. 11, 1961.* Plenum Press, New York.

Albers, V.M., 1965. *Underwater Acoustics Handbook II.* Pennsylvania State University Press, University Park, Pa.

Bartberger, C.L., 1956. *Lecture Notes on Underwater Acoustics.* U.S. Naval Air Development Center, Johnsville, Pa.

Bobber, R.J., 1970. *Underwater Electroacoustic Measurements.* U.S. Government Printing Office, Washington, D.C.

Brekhovskikh, L.M., 1974. *Acoustics of the Ocean.* Izdatel'stvo Nauka, Moscow.

Horton, J.W., 1957. *Fundamentals of Sonar.* U.S. Naval Inst., Annapolis, Maryland.

Kinsler, L.K. and Frey, A.R., 1962. *Fundamentals of Acoustics.* Wiley, New York.

Officer, C.B., 1958. *Introduction to the Theory of Sound Transmission with Application to the Ocean.* McGraw-Hill, New York.

Tolstoy, I. and Clay, C.S., 1966. *Ocean Acoustics: Theory and Experiments in Underwater Sound.* McGraw-Hill, New York.

Tucker, D.G. and Glazey, B.K., 1966. *Applied Underwater Acoustics.* Pergamon Press, London.

Urick, R.J., 1967. *Principles of Underwater Sound for Engineers.* McGraw-Hill, New York.

Problem 1.6

The mean square pressure is calculated from:

$$\overline{p^2} = \overline{p_1^2 \sin^2 \omega_1 t} + \overline{p_2^2 \sin^2 \omega_2 t} + \overline{2p_1p_2 \sin \omega_1 t \sin \omega_2 t}$$

$$\overline{p^2} = p_1^2 \overline{\sin^2 \omega_1 t} + p_2^2 \overline{\sin^2 \omega_2 t}$$

$$\overline{p^2} = \tfrac{1}{2}(p_1^2 + p_2^2)$$

This value is inserted into equation 1.9 to get intensity. The general form of the mean square pressure is:

$$\overline{p^2} = \frac{1}{2} \sum_{i=1}^{n} p_i^2$$

Problem 1.7

The mean square pressure is calculated from:

$$\overline{p^2} = \frac{p_0^2}{T} \int_0^T \exp\left(-\frac{2t}{\tau_0}\right) \cos^2 \omega t \, dt$$

$$\overline{p^2} = \frac{p_0^2}{2T}\left[\int_0^T \exp\left(-\frac{2t}{\tau_0}\right) dt + \int_0^T \exp\left(-\frac{2t}{\tau_0}\right) \cos 2\omega t \, dt\right]$$

$$\overline{p^2} = \frac{p_0^2}{2T}\left[1 - \exp\left(-\frac{2t}{\tau_0}\right)\right]\left[\frac{2 + \omega^2\tau_0^2}{1 + \omega^2\tau_0^2}\right]$$

Letting $\omega = 2\pi/T$, $\alpha = \tau_0/T$, then:

$$\overline{p^2} = \frac{p_0^2\alpha}{4}\left[1 - \exp\left(-\frac{2}{\alpha}\right)\right]\left[\frac{2 + 4\pi^2\alpha^2}{1 + 4\pi^2\alpha^2}\right]$$

When $T \gg \tau_0$, $\alpha \to 0$ and $\overline{p^2} = 1/2\, p_0^2\alpha$.
When $\tau_0 \gg T$, $\alpha \to \infty$ and $\overline{p^2} = 1/2\, p_0^2$.

Problem 1.8

In one dimension the wave equation becomes:

$$\frac{\partial^2 p}{\partial x^2} - \frac{1}{c^2}\frac{\partial^2 p}{\partial t^2} = 0$$

Let $z = kx - \omega t$, then:

$$\frac{\partial p}{\partial x} = \frac{\mathrm{d}p}{\mathrm{d}z}\frac{\partial z}{\partial x} = kf'$$

$$\frac{\partial^2 p}{\partial x^2} = k^2 f''$$

$$\frac{\partial p}{\partial t} = -\omega f'$$

and:

$$\frac{\partial^2 p}{\partial t^2} = \omega^2 f''$$

The wave equation reduces to:

$$k^2 f'' - \frac{\omega^2}{c^2} f'' = 0$$

Therefore $p = f(kx - \omega t)$ is a solution of the wave equation and $c^2 = \omega^2/k^2$.

Problem 1.9

In one dimension the equation becomes:

$$\frac{\partial p}{\partial x} = -\rho\frac{\partial u}{\partial t}$$

Using the parameter z, c, and the differentials as in Problem 1.8, the equation reduces to:

$$k\frac{\mathrm{d}p}{\mathrm{d}z} = \rho\omega\frac{\mathrm{d}u}{\mathrm{d}z}$$

$$\frac{\mathrm{d}p}{\mathrm{d}z} = \rho c\frac{\mathrm{d}u}{\mathrm{d}z}$$

Integrating over z and discarding constant terms results in $p = \rho cu$.

Problem 2.1

(a) $\omega_{\mathrm{res}} = [L_{\mathrm{M}} C_{\mathrm{M}}]^{-1/2} = [0.107 \times 0.069 \cdot 10^{-6}]^{-1/2}$

$= 1.16 \cdot 10^4$ rad/sec

$f_{\mathrm{res}} = \omega_{\mathrm{res}}/2\pi = 1.83 \cdot 10^3$ Hz

(b) $Q_{\mathrm{M}} = \omega_{\mathrm{res}} L_{\mathrm{M}}/(R_{\mathrm{M}} + R_{\mathrm{R}})$

$= 0.107 \times 1.16 \cdot 10^4/1150 = 10.8$

(c) $Q_e = \omega_{res} C_0 (R_M + R_R)$

$= 1.16 \times 10^4 \times 0.069 \cdot 10^{-6} \times 1150$

$= 0.92$

(d) $\Delta f = f_{res}/Q_M = 1850/10.8 = 170\,\text{Hz}$

(e) $Q = (Q_M + Q_e)/2 = 11.7/2 = 5.8$

$f = 1850/5.8 = 320\,\text{Hz}$

(g) $$k = \left[\frac{\pi^2}{8}\frac{C_M}{C_0}\right]^{1/2} = \left[\frac{\pi^2 \times 2680 \cdot 10^{-12}}{8 \times 0.069 \cdot 10^{-6}}\right]^{1/2}$$

$= 0.22$

Problem 2.2

Admittance is given by:

$$Y_M = (Z_M + Z_R)^{-1}$$

Here $Z_R = R_R$ and $Z_M = R_M + jX_M$. Putting this into the admittance equation results in:

$$Y_M = \frac{1}{(R_R + R_M) + jX_M} = \frac{1}{R + jX_M}$$

$$= \frac{R}{R^2 + X_M^2} - \frac{jX_M}{R^2 + X_M^2}$$

$$= G_M + jB_M$$

Where conductance (G_M), susceptance (B_M), resistance (R), and reactance (X_M) are given by:

$$G_M = R/(R^2 + X_M^2)$$

$$B_M = -X_M/(R^2 + X_M^2)$$

$$R = R_R + R_M$$

$$X_M = \omega L_M - \frac{1}{\omega C_M}$$

These equations describe a circle of radius $1/R$ in the complex admittance plane (G_M-abscissas, jB_M-ordinate). The following table gives some of the points in the plane.

ω	X_M	G_M	B_M
0	$-\infty$	0	0
∞	$+\infty$	0	0
ω_{res}	0	$1/R$	0
ω_1	$X_M = -R$	$1/2R$	$1/2R$
ω_2	$X_M = +R$	$1/2R$	$-1/2R$

Problem 2.4

The quantity $r^2 p_0^2 t_0$ is proportional to the total energy flux through any surface enclosing the source and should be independent of r and directly proportional to the charge weight W. First to show that it is proportional to total energy. The total energy is:

$$T = \int_\sigma E\,\mathrm{d}s$$

where σ is a closed surface about the source. Assume that it is a sphere of radius r. Then:

$$T = \frac{1}{2\rho c}\int_\sigma p_0^2 t_0 r^2 \sin\theta\,\mathrm{d}\theta\,\mathrm{d}\phi = \frac{4\pi}{2\rho c} r^2 p_0^2 t_0$$

Finally we evaluate the parameter:

$$r^2 p_0^2 t_0 \propto r^2 \left[\left(\frac{W^{1/3}}{r}\right)^{1.13}\right]^2 \left(\frac{W^{1/3}}{r}\right)^{-0.22} W^{1/3}$$

$$\propto r^{-0.04} W^{1.01}$$

$$\propto W[r^{-0.04} W^{0.01}]$$

Problem 3.2

For a spherically symmetric field and disregarding the harmonic time factors, the pressure and velocity are:

$$p(r) = \frac{A}{r}\exp(jkr)$$

and:

$$u(r) = \frac{A}{\rho c r}\left(1 - \frac{j}{kr}\right)\exp(jkr)$$

respectively. Volume velocity is the surface integral of the source's surface velocity, $u(a)$, over the source's surface, $r = a$. For a sphere vibrating in its fundamental mode the volume velocity is:

$$Q = \int_{\Sigma} u(a)\,\mathrm{d}s = u(a)\int_{4\pi} a^2 \mathrm{d}\Omega = 4\pi a^2 u(a)$$

After substituting the surface velocity, we get:

$$Q = \frac{4\pi a A}{\rho c}\left(1 - \frac{j}{ka}\right) \exp\,(jka)$$

Its magnitude is:

$$|Q| = \frac{4\pi a\,|A|}{\rho c}\left(1 + \frac{1}{k^2 a^2}\right)$$

Dividing the magnitude of the pressure by the magnitude of the volume velocity gives:

$$\frac{|p(r)|}{|Q|} = \frac{\rho c}{4\pi a r}\left(1 + \frac{1}{k^2 a^2}\right)^{-1}$$

For a very small source, that is for $ka \ll 1$:

$$\frac{|p(r)|}{|Q|} = \frac{\rho c k}{4\pi r}$$

Combining equations 3.6 and 3.7, we find:

$$\frac{|p(r_0)|}{|Q|} = \frac{h(f)}{g(r_0)} = \frac{10^{-7}}{J_s} = \frac{\rho c k}{4\pi r_0} = \frac{\rho c}{2\lambda r_0}$$

At $r_0 = 1$ meter:

$$J_s = \frac{2\lambda}{\rho c} 10^{-7}$$

Problem 3.4

To show that:

$$\sum_{m=0}^{n} \exp(jmu) = \frac{\exp(jNu) - 1}{\exp(ju) - 1} = \exp\left[\frac{j(N-1)u}{2}\right] \frac{\sin\,(Nu/2)}{\sin\,(u/2)}$$

Multiply and divide left-hand side by $[\exp\,(ju) - 1]$ to get:

$$\frac{[\exp(ju) - 1]\ \Sigma \exp\,(jmu)}{\exp\,(ju) - 1} = \frac{\Sigma \exp\,[j(m+1)u] - \Sigma \exp\,(jmu)}{\exp\,(ju) - 1}$$

This reduces to:

$$\Sigma \exp\,(jmu) = \frac{\exp\,(jNu) - 1}{\exp\,(ju) - 1}$$

Factor out exp $(jNu/2)$ in the numerator and exp $(ju/2)$ in the denominator to get:

$$\Sigma \exp(jmu) = \frac{\exp\left(\frac{jNu}{2}\right)\left[\exp\left(\frac{jNu}{2}\right) - \exp\left(-\frac{jNu}{2}\right)\right]}{\exp\left(\frac{ju}{2}\right)\left[\exp\left(\frac{ju}{2}\right) - \exp\left(-\frac{ju}{2}\right)\right]}$$

$$= \exp\left[\frac{j(N-1)u}{2}\right]\frac{\sin(Nu/2)}{\sin(u/2)}$$

Problem 4.1

Array gain is given by equation 4.5 where $\overline{s^2}$ and $\overline{n^2}$ are subject to equation 4.4. For an isotropic field, the noise integral in equation 4.4 becomes $\overline{N^2} = 4\pi\overline{n^2} = \overline{S^2}$. A unidirectional signal may be represented in terms of a Dirac delta function (see p. 75), that is, $\overline{s^2} = \overline{S^2}\,\delta(0, 0)$. The signal integral in equation 4.4 becomes:

$$\int \overline{s^2}\, d\Omega = \overline{S^2}\int \delta(0, 0)\, d\Omega = \overline{S^2}$$

Therefore, the array gain becomes:

$$AG = 10 \log \frac{\overline{s^2} \int \delta(0, 0)\, b(\theta, \phi)\, d\Omega}{\frac{\overline{S^2}}{4\pi}\int b(\theta, \phi)\, d\Omega}$$

This reduces to quickly to directivity index when the property (3) equation 6.26 page 75 of the Dirac delta function is used.

Problem 6.1

The surface integrals of equation 6.28 are zero. Substituting for ψ and G into the volume integral gives (with $p \equiv \phi$):

$$p(r) = \iiint_V \frac{\exp[jk|r - r'|]}{|r - r'|}\{a\delta(r' - r_0)\}\, dV'$$

This reduces to:

$$p(r) = \frac{a \exp[jk|r - r_0|]}{|r - r_0'|} = \frac{a \exp[jkR]}{R}$$

where $R = |r - r_0|$.

Problem 6.2

Consider a point r'_+ on the positive side of the plane and its image point r'_- on the negative side. The Green's function which is zero on the plane is:

$$G(r, r') = \frac{\exp[jk|r - r'_+|]}{|r - r'_+|} - \frac{\exp[jk|r - r'_-|]}{|r - r'_-|}$$

Substituting this Green's function and the given ψ into the volume integral of equation 6.28 and noting that the surface integrals are zero (because the pressure field and the Green's function are zero on the surface) gives:

$$p(r) = a \iiint_V \left\{ \frac{\exp[jk|r - r'_+|]}{|r - r'_+|} - \frac{\exp[jk|r - r'_-|]}{|r - r'_-|} \right\} \delta(r - r'_0)\, dV'$$

Note that:

$$r'_+ = r' = (x', y', z')$$

and:

$$r'_- = (x', y', -z')$$

then:

$$r'_- = (x', y', z') - 2z'1_z = r' - 2z'1_z$$

where 1_z is a unit vector in the z-direction. Substituting this into the volume integral gives:

$$p(r) = a \iiint_V \left\{ \frac{\exp[jk|r - r'|]}{|r - r'|} - \frac{\exp[jk|r - r' + 2z'1_z|]}{|r - r' + 2z'1_z|} \right\} \cdot \delta(r' - r_0)\, dV'$$

The results of the integration are:

$$p(r) = \frac{a \exp[jk|r - r_0|]}{|r - r_0|} - \frac{a \exp[jk|r - r_0 + 2z_0 1_z|]}{|r - r_0 + 2z_0 1_z|}$$

This can be further reduced by setting:

$$|r - r_0| = [x^2 + y^2 + (z - z_0)^2]^{1/2} = R_+$$

and:

$$|r - r_0| = [x^2 + y^2 + (z - z_0 + 2z_0)^2]^{1/2}$$

$$= [x^2 + y^2 + (z + z_0)^2]^{1/2} = R_-$$

Then:

$$p(r) = \frac{a \exp jkR_+}{R_+} - \frac{a \exp jkR_-}{R_-}$$

As was probably obvious from the start the field is the sum of that produced by the object and its image.

Problem 10.1

The input signal is:

$$x(t) = a(t)\cos(\omega t + \phi) + n(t)$$

The in-phase signals are given by:

$$I_1 = a(t)\cos\omega t \cdot \cos(\omega t + \phi) + n(t)\cos\omega t$$

$$= \tfrac{1}{2}a(t)\cos\phi + \tfrac{1}{2}a(t)\cos(2\omega t + \phi) + n(t)\cos\omega t$$

$$I_2 = a(t)\cos\phi + \text{small noise terms}$$

$$I_3 = \cos^2\phi \int_0^T a^2(t)\,dt + \text{s.n.t.}$$

Similarly for the quadrature signals:

$$Q_3 = \sin^2\phi \int_0^T a^2(t)\,dt + \text{s.n.t.}$$

And the output is:

$$\hat{\rho} = [I_3 + Q_3]^{1/2} = \int_0^T a^2(t)\,dt + \text{s.n.t.}$$

Problem 10.2

The probability of detection given hypothesis H_1 and threshold setting α_0 is given by:

$$P_D = \frac{1}{\sqrt{2\pi}\,\sigma}\int_{\alpha_0}^{\infty} \exp\left\{-\frac{1}{2}\left(\frac{\alpha - \mu_{sn}}{\sigma}\right)^2\right\} d\alpha$$

Using the definition of detection index (d), the probability of detection can be written:

$$P_D = \frac{1}{\sqrt{2\pi}\,\sigma}\int_{\alpha_0}^{\infty} \exp\left[-\tfrac{1}{2}(t_0 - \sqrt{d})^2\right] d\alpha$$

where $t_0 = (\alpha_0 - \mu_n)/\sigma$.

Letting $x = t_0 - \sqrt{d}$, then:

$$P_D = \frac{1}{\sqrt{2\pi}}\int_{t_0 - \sqrt{d}}^{\infty} \exp\left(-\tfrac{1}{2}x^2\right) dx$$

Then P_D is given by the complementary error function:

$$P_D = \tfrac{1}{2}\,\mathrm{erfc}\left(\frac{t_0 - \sqrt{d}}{\sqrt{2}}\right)$$

Similarly the probability of false alarm can be derived from:

$$P_{\mathrm{FA}} = \frac{1}{\sqrt{2\pi}\,\sigma}\int_{\alpha_0}^{\infty} \exp\left[-\frac{1}{2}\left(\frac{\alpha-\mu_{\mathrm{n}}}{\sigma}\right)^2\right] \mathrm{d}\alpha$$

INDEX